Pocket Guide to Biomolecular NMR

Michaeleen Doucleff ·
Mary Hatcher-Skeers · Nicole J. Crane

Pocket Guide to Biomolecular NMR

Dr. Michaeleen Doucleff
National Institutes of Health (NIH)
National Institute of Diabetes &
Digestive & Kidney Deseases
(NIDDK)
Buildg.5
Center Drive 31
20892-0510 Bethesda Maryland
USA
mcdoucleff@gmail.com

Dr. Mary Hatcher-Skeers
The Claremont Colleges
WM Keck Science Center
North Mills Avenue 925
91711 Claremont California
USA
mhatcher@jsd.claremont.edu

Dr. Nicole J. Crane
Henry M. Jackson Foundation for the
Advancement of Military Medicine
Naval Medical Research Center
Dept. Regenerative Medicine
Robert Grant Avenue 503
20910 Silver Spring Maryland
USA
Nicole.Crane@med.navy.mil

ISBN 978-3-642-16250-3 e-ISBN 978-3-642-16251-0
DOI 10.1007/978-3-642-16251-0
Springer Heidelberg Dordrecht London New York

Cover design: WMXDesign GmbH, Heidelberg

Printed on acid-free paper

Springer is part of Springer Science+Business Media (www.springer.com)

Preface

> *...energy is applied to the sample in the form of short, intense pulses, and nuclear signals are observed after the pulses are removed. The effects which result can be compared to the free vibration or "ringing" of a bell...*
>
> Eric Hahn, 1953, Physics Today 6, 4–9

The essence of NMR is quite simple. You place a sample in a magnetic field, apply another field, and then collect the electromagnetic wave generated by the sample. No need to ionize, crystallize, burn, or even heat up the sample. A plain-vanilla solution of a compound in water is all you need to identify the substance and characterize many of its key biophysical properties.

Unfortunately, the elegance and beauty of NMR is often lost when students open a textbook. Even the simplest introductions on NMR immediately immerse students in mathematics and theories laced with constants and concepts that lack tangible, macroscopic manifestations. For example, many textbooks begin by defining the energy of a nucleus in a magnetic field, $E = -\gamma\hbar m B_o$, a simple equation that is clear and understandable to all. However, take a closer look at this equality. Hidden inside the constants and variables are sophisticated concepts that are tough for many of us to relate to. Take "γ" for instance. Wikipedia defines "γ" (i.e., the gyromagnetic ratio) as the ratio of the nucleus's magnetic dipole moment to its angular momentum. Riding the swings at a carnival definitely helps us to relate to angular momentum. But what exactly is a magnetic dipole moment, and why would taking the ratio of the magnetic dipole moment and the angular moment be proportional energy? Very quickly, the theory and mathematics behind NMR plunges into ideas and phenomenon that aren't present in our macroscopic world, making them difficult to grasp, remember, and use.

In contrast to the labyrinth of equations in NMR textbooks, performing actual NMR experiments is remarkably easy. You place the sample in the magnet, set a couple of parameters, and hit "go." After a few commands at the keyboard, your spectrum appears on the screen with a peak for each hydrogen in the compound. Can't get much easier than that! So why can't the theory of NMR be that simple too?

That, indeed, is the purpose of this book—to bring the common sense and practicality of experimental NMR to the theory of NMR. To accomplish this task, we set three major goals:

1. *Preserve the simplicity of NMR:* Instead of using equations and complex formulism to explain *why* nuclei behave as they do in a magnetic field, we simply describe *how* nuclei act during NMR experiments. We minimize jargon and keep the descriptions short, pithy, and easy to understand. Students with minimal experience in chemistry and biology should easily follow the entire book. The only equations used in the main text are the sine and exponential functions.
2. *Add a hefty dose of intuition to NMR:* In place of mathematics and formulism, we use concrete analogies to which readers can relate, making concepts easy to assimilate and use. In this regard, we have created a new framework for explaining NMR experiments. Instead of trying to define "spin," we simply state that nuclei "ring" in magnetic fields like tiny bells. This bell analogy, which is employed throughout the book, has never been used to explain NMR and makes it surprisingly easy to learn advanced NMR concepts, such as dipole–dipole coupling and relaxation theory.
3. *Flatten the learning curve:* The "bell" analogy also provides a new language for discussing NMR experiments. Because this language is based on an intuitive model of NMR, students quickly master

it. In other words, this new approach flattens the steep learning curve of NMR and makes NMR accessible to students at all levels, even those with little experience in spectroscopy, quantum mechanics, or physics. Furthermore, the intuitive perspective presented in this book will help advanced students remember and integrate more mathematical explanations of NMR into their experimental designs and analyses.

The small size and fast pace of the book makes it well suited as a companion to traditional biophysics or biochemistry textbooks at the undergraduate or graduate levels. However, we hope that the book will also be useful for professional researchers in the biological sciences who are interested in collaborating with NMR spectroscopists or in garnering a better understanding of research articles that present NMR data.

Finally, one note about the style and tone of the book. The book is written to be read. Therefore, we use a light, fun style that will hopefully hold your attention from the initial descriptions of one-dimensional experiments to the final pages covering CPMG sequences. We hope that you enjoy it!

Michaeleen Doucleff
Mary Hatcher-Skeers
Nicole J. Crane

Contents

Chapter 1
Atomic Bells and Frequency Finders

1.1 Chemical Choirs

Have you ever heard a handbell choir? Often associated with Christmas, the resonating sound of a handbell choir evokes images of ancient cathedrals and snowy nights. But a handbell choir is just as much a feast for the eyes as it is for the ears.

Handbell choirs contain 12 members, each handling 2–12 bells for a total of more than 50 bells. The shape and size of each bell is precisely tuned to resonate at exactly one note. The largest bells, which have diameters greater than a foot, play the bass notes with low frequencies and long wavelengths (Fig. 1.1), and the smallest bells, which weigh only a few ounces, play the soprano notes with higher frequencies and short wavelengths. Together the set of bells covers over seven octaves.

Molecules are similar to bell choirs. The nucleus of each atom "rings" at its own frequency depending on its size, shape, and neighboring atoms. Like handbells, the larger atoms ring at lower frequencies and the smaller atoms ring at higher frequencies. For example, the tiny hydrogen atoms ring with frequencies approximately ten times greater than the nitrogen and carbon atoms (Fig. 1.2). So the hydrogens are definitely the sopranos, bolting out the high frequency notes, while the carbon and the nitrogens are more like the basses, softly providing the low notes.

M. Doucleff et al., *Pocket Guide to Biomolecular NMR*,

DOI 10.1007/978-3-642-16251-0_1,

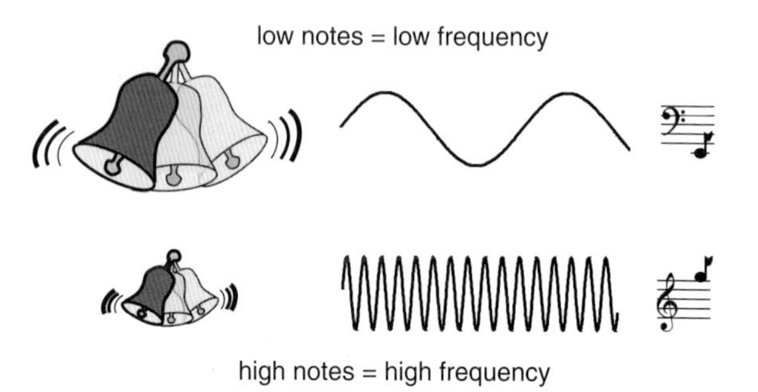

Fig. 1.1 In a bell choir, large bells play low notes with small frequencies (*top*), whereas smaller bells play higher notes with larger frequencies (*bottom*)

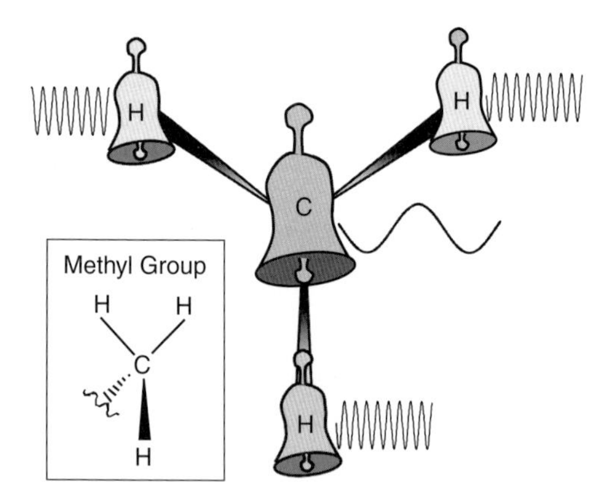

Fig. 1.2 Atoms in molecules are like bells in the choir: the smaller hydrogen atoms in the methyl group ring at higher frequencies than the larger carbon (or nitrogen) atoms

The difference between bells in a choir and atoms in a molecule is the type of wave each produces. Bells create sound waves or oscillating changes in the density of air molecules. Nuclei create electromagnetic waves or oscillating changes in the strength of the electromagnetic field. For bells, the ringing frequency depends on the bell's size and the material it is made of. For example, big bells create low-frequency notes (Fig. 1.1), similar to how the thumbing bass of techno is projected by big, imposing speakers. For atoms, the ringing frequency depends primarily on two items:

(a) The type of atom;
(b) The strength of the magnetic field around the atoms (hence the electroMAGNETIC waves they produce).

$$\text{ringing frequency} = (\text{type of atom}) \times (\text{magnetic field strength})$$

The larger the magnetic field, the faster the ringing frequency. For example, in a magnetic field of 3 T, the field strength of modern MRI machines used to image soft tissue, hydrogen atoms in water ring at 128 million times per second or 128 MHz. To put this frequency in perspective, if these were sound waves, 128 MHz would be too fast for the human ear to detect; we can hear sound only between 20 and 20,000 Hz, or 100-times slower than the frequency of hydrogen nuclei. Thus, these atomic bells ring extraordinarily fast in a 3-T magnetic field. But in the Earth's magnetic field (0.00005 T), hydrogens in water ring at approximately 2,000 Hz, an audible frequency for sound waves.

Remember, these are electromagnetic waves, like the ones broadcasting music to your radio and sitcoms to your network television. In fact, many modern NMR machines operate around 11.6 T or approximately 500 MHz, and 512–608 MHz are reserved for television channels 12–36 (in the USA). Next time you are watching these channels, imagine how all the hydrogen atoms in your body would emit electromagnetic waves just like the ones broadcast by NBC if you were lying in an 11.6 T (500 MHz) NMR machine.

1.2 Essentials of Electromagnetism

Since we are dealing with electromagnetic waves, let's take a moment to learn a bit more about the phenomenon called "electromagnetism."

Look up electromagnetism in any physics textbook, and you'll be flooded with equations and vague terminology. Luckily, these are not necessary for a practical understanding of biomolecular NMR. Instead, let's focus on a few key descriptive ideas of electromagnetism that will take you surprisingly far while reading biomolecular NMR papers and textbooks.

We all know that electrons create a force field that attracts positively charged particles, such as protons. When the electron is standing still, this negative force field is called an "electrical field," and when the electron moves, it creates a "magnetic field." Together the electric and magnetic fields constitute the "electromagnetic field." That is almost all you that you need to remember for biomolecular NMR: When electrons sit still, they create electric fields; when electrons move, they create magnetic fields. And, the magnetic field runs perpendicular to the movement of the electrons. So if your laptop's electrical cord is lying flat on a table, the magnetic field created by the current in the wire forms circles (concentric circles) around the wire, perpendicular to the river of electrons flowing in the wire (Fig. 1.3).

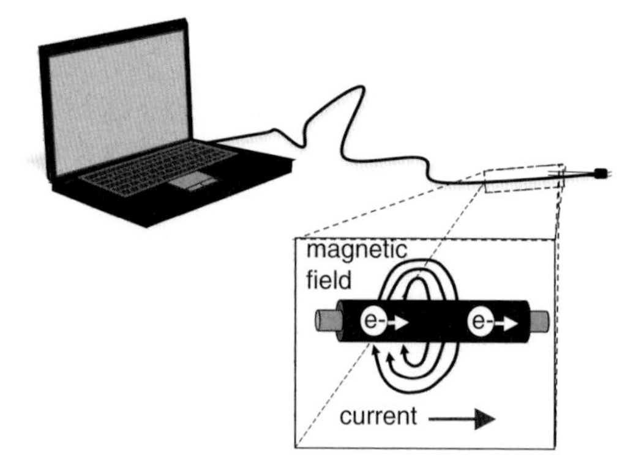

Fig. 1.3 Traveling through the electric cord of you laptop, electrons create a magnetic field that forms concentric circles around the cord, perpendicular to the electrons' motion (the current)

1.3 Electromagnetic Microsensors

Okay, so atoms create electromagnetic waves or "ring" at particular frequencies—what good is that? The key, already alluded to in point *b* in Sect. 1.1, is that the atom's frequency depends on the strength of its surrounding magnetic field. Let's look at an analogy to see how this works.

One reason Napa Valley is a great winemaking region is its microclimates. The southern valley, which receives cool breezes and fog from the San Francisco and San Pablo bays, has an average temperature 10–15°F lower than those in the central part of Napa Valley merely 15 miles north (Fig. 1.4). More microclimates are created by the hills surrounding the valley (Fig. 1.4), and each grape variety responds differently to these subtle climate changes, producing interesting flavors and nuances in the wine. For example, on the hot valley floor, the Cabernet Sauvignon grapes fully ripen with a sweet, fruity flavor,

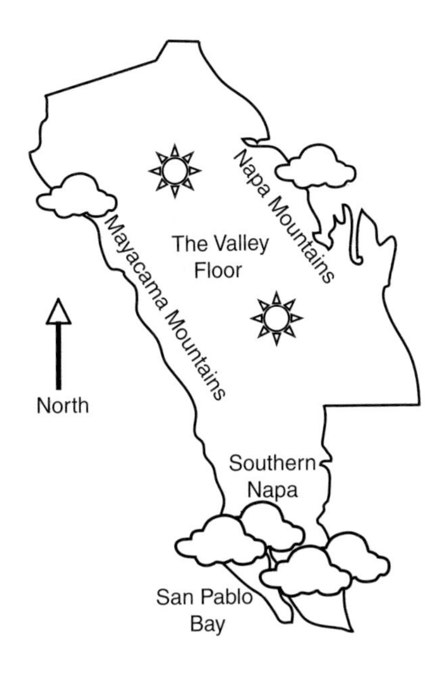

Fig. 1.4 A map of Napa Valley showing how the clouds from the San Pablo Bay and the surrounding mountains create microclimates around the valley

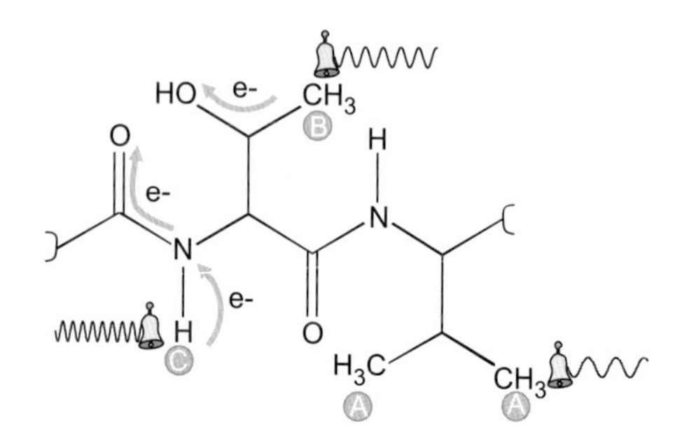

Fig. 1.5 A nucleus's ringing frequency depends on the electromagnetic environment: When nearby atoms, such as oxygen and nitrogen atoms, strip away electrons, the nucleus rings faster than nuclei that are not close to these electronegative atoms

but on the surrounding hillside at 2,000 ft above sea level, where the evenings are slightly cooler, the Cabernet Sauvignon grapes keep more of their acidity, giving them a tart or slightly sour flavor. Thus, the grapes are microclimate *sensors* that detect and respond to the slight weather variations in their immediate surroundings.

Atomic nuclei are microsensors like the grapes, but instead of sensing subtle variations in weather, nuclei sense tiny changes in their surrounding electromagnetic field. Instead of responding by varying their flavor, the nuclei respond by ringing at different frequencies.

Let's take an amino acid in a peptide as an example (Fig. 1.5). The methyl hydrogens at the end of the valine side chain (A) feel a significantly different electromagnetic environment than the methyl hydrogens at the end of the threonine side chain (B). Why? Sitting in the 11.75-T magnet, these two hydrogen nuclei feel approximately the same magnetic field, which causes them to ring at approximately 500 MHz (500,000,000 Hz). (That's why NMR spectroscopists call an 11.75-T magnet, a "500 MHz" magnet, even though MHz is a unit of frequency and *not* magnetic field.)[1] But the nuclei in a molecule are not alone. They are surrounded by quickly moving electrons, and, as we have just learned, when electrons move, they create their own magnetic fields. These "micro-magnetic fields" in compounds slightly change the ringing frequency of the atoms, just like the "micro-climates" of Napa Valley slightly change the flavor of its wines.

It turns out that the electrons moving around nuclei tend to oppose or cancel out the external magnetic field. Therefore, the total magnetic field felt by a hydrogen nucleus is the 500 MHz of the external field minus the micro-magnetic fields created by orbiting electrons. So the *less* electrons nearby, the *stronger* the magnetic field felt by a nucleus and the faster its nucleus rings (remember point *b* in Sect. 1.1). In other words, a partially "naked" nucleus stripped of its electrons by neighboring atoms rings *faster* than a nucleus located in a high density of electrons, which shields the nucleus from the strong external magnetic field (skip ahead to Fig. 4.5 to see a beautiful example of this).

Oxygen loves electrons. It sucks away electrons from all neighboring atoms. So, a hydrogen near an oxygen will ring faster than a hydrogen surrounded by less electron-sucking (or "electronegative") atoms, like carbon and other hydrogens. Indeed, in the peptide of Fig. 1.5, nucleus B near the oxygen of the threonine rings at an approximately 125 Hz higher frequency than methyl hydrogens of the valine residue (nuclei A), which are far away from the electron-stripping oxygen.

The most important hydrogens for biomolecular NMR spectroscopists are those attached to the nitrogens in peptide bonds (atom C in Fig. 1.5). We call these hydrogens "amide protons," and they ring relatively *fast* for nuclei in proteins, with frequencies ~2,000 Hz faster than most of the other hydrogens in a protein. This large frequency difference is primarily due to the electronegativity of

[1] Note, NMR spectroscopists rarely use the units "Teslas" to describe the field strength of a spectrometer; instead they refer to magnetic field strengths by the frequency at which the hydrogens ring; for example, 500 MHz for an 11.75 T magnet.

the attached nitrogen and the neighboring peptide bond, which are both electron sinks, stripping electrons away from the amide proton and exposing it to the strong external magnetic field (atom C in Fig. 1.5). And, remember, the stronger the magnetic field, the faster the atoms ring.

1.4 Frequency Finders

In a nutshell, an NMR experiment is like a big hammer that rings the atoms' nuclei and then listens for the electromagnetic waves they create (Fig. 1.6). This is pretty straightforward for a small methyl group with only three atoms as in Fig. 1.6, but in biology we work with proteins and nucleic acids with hundreds of nuclei all ringing at the same time. Plus, frequency differences between these nuclei are quite small. In the peptide described above, nuclei A and B ring with frequencies that are only 175 Hz apart—that's only a 0.000035% difference in frequency! The result is a chemical cacophony that's not much use for us biochemists in this raw form (Fig. 1.7a).

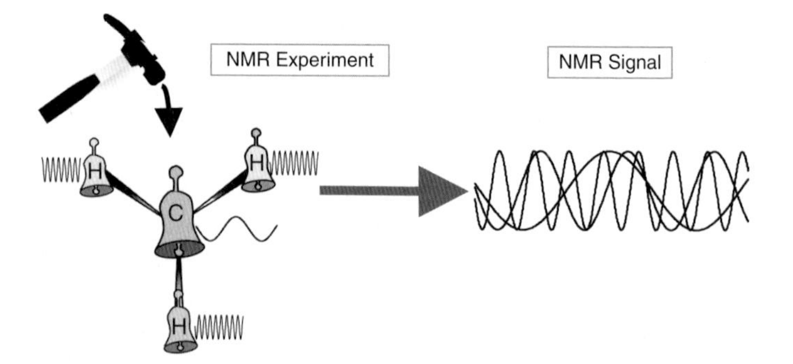

Fig. 1.6 An NMR experiment is like a "big hammer" that rings all the nuclei in a molecule and then listens to the electromagnetic waves they create

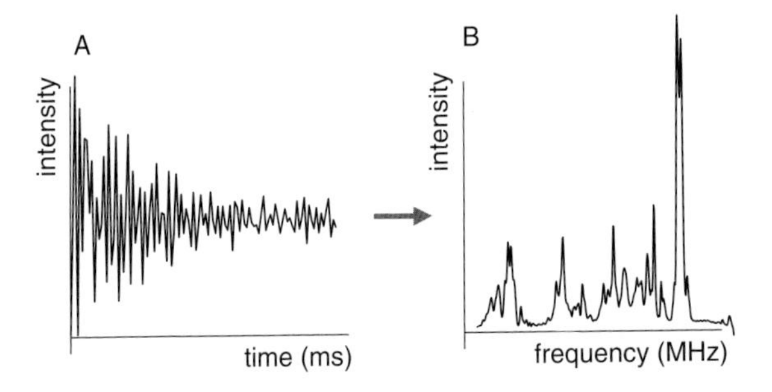

Fig. 1.7 (**a**) Dozens of different nuclei ringing simultaneously produce dozens of sine waves with different frequencies piled on top of each other. (**b**) The Fourier Transform (FT) converts this cacophonous signal into a useful plot by changing the x-axis from time (ms) to frequency (MHz) with one peak for each type of sine curve

Our NMR data is a collection of electromagnetic sine curves piled on top of each other (Fig. 1.7a). Specifically, it has one sine curve for each ringing nucleus. To study and characterize specific atoms in a protein or nucleic acid, we need to extract out each atom's ringing frequency from this molecular rock concert. That's where brilliant mathematicians and computer scientists come to our rescue. With a flick of the keyboard, we can take what looks like an electromagnetic mess (Fig. 1.7a) and convert it into a graph that's easy to read and understand (Fig. 1.7b). Notice that the only difference between the graphs in a and b in Fig. 1.7 is the x-axis: In a, the x-axis has units of time (seconds), but in b, the units are frequency (1 Hz = 1 per second).

The magic function that converts the x-axis into frequency is called the "Fourier Transform," and it is extremely powerful. Although the Fourier Transform seems a bit complicated at first glance, in fact, it is surprisingly easy to understand. The Fourier Transform takes a curve, such as the one shown in Fig. 1.7a, and simply finds the frequency of each sine wave present in that curve. If the curve contains a sine wave with frequency 500.123 MHz, then the Fourier Transform gives a positive value at the frequency of 500.123 MHz. If the sine wave is missing in the curve, then the Fourier Transform gives zero for that frequency. That's it! The Fourier Transform is merely a frequency finder that gives a "peak" for each frequency seen in a curve. When NMR spectroscopists say "we apply a Fourier Transfer" or "FT" for

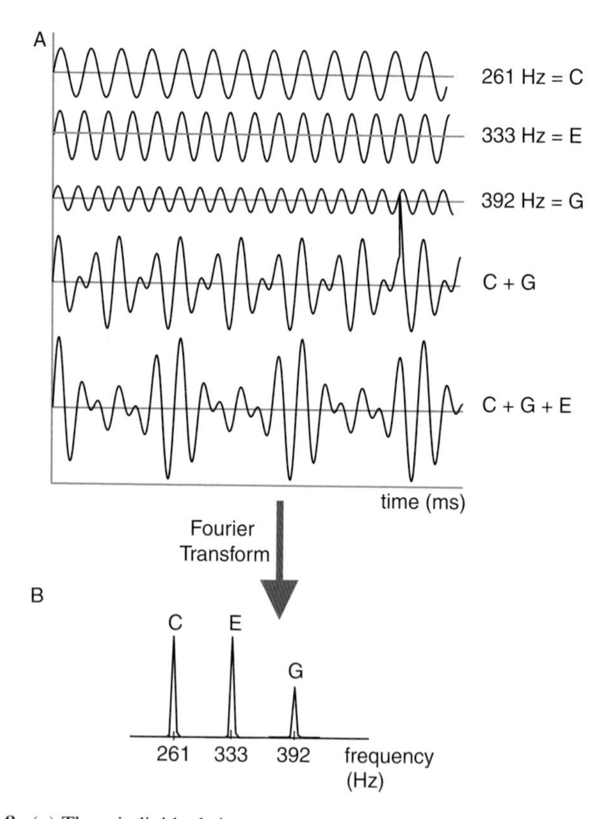

Fig. 1.8 (**a**) Three individual sine curves, those for the notes C, E, and G, combine together to create a major chord; (**b**) the Fourier Transform of this combined curve produces three peaks, one for each note in the chord

short, all they are doing is finding the frequencies of every sine curve present in their data.

Let's explore the Fourier Transform further with a real-world example. Crosby, Stills, and Nash were a groovy band back in the 1970s, but their beautiful harmonies are still appreciated today. When the old-school band sings in perfect harmony, three sine waves combine together to create the chord that you hear (Fig. 1.8a). If we apply a Fourier Transform to the sound wave, we would find three peaks in a plot of intensity versus frequency (Fig. 1.8b)—one peak at the low frequency of Crosby's voice (261 Hz = "middle C"), one at a higher frequency of Stills' voice (333 Hz = "E"), and the third at the highest frequency for the falsetto note of Nash's voice (392 Hz = "G"). In addition, the height of the peak indicates how loud the note is sung. As shown here (Fig. 1.8b), Crosby's high note is half as loud as the other two notes.

How does the Fourier Transform work? Instead of giving a mathematical explanation (which is shown in the Mathematical Sidebar 1.1), let's use an analogy that will also help us understand other concepts presented in later chapters. Imagine a relay event in short track speed skating at the Winter Olympics. In this sport, one team member starts out racing around the track while a second teammate begins skating slowly around the inside of the track. When it's time for the second team member to enter the race, she speeds up and then gets a big push from the first teammate to help "jump start" the second leg of the rely. If the two teammates are at the same location at the same time on the track, then the first racer can provide a strong push to accelerate the new racer. However, if the two racers are not at the exact same location, the push is weak, and the second racer misses out on the speed boost.

If we plot the racer's horizontal position on the track with respect to time, we would get a curve similar to our NMR data (Fig. 1.9a). To maximize contact between the teammates (and thus the speed boost), the teammates' frequencies around the track must match. The second racer could find this "sweet spot" by testing a whole range of frequencies around the track—slow, medium, or fast pace—just like

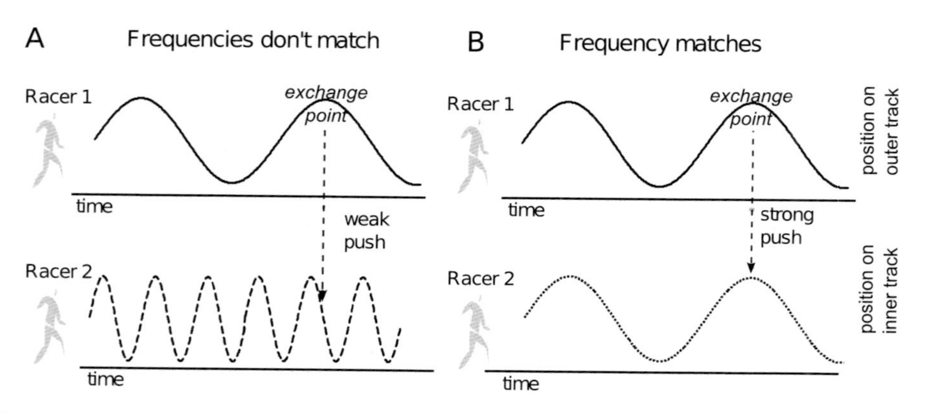

Fig. 1.9 The Fourier Transform performs a type of frequency matching similar to the one shown above for two teammates in a speed skating relay at the Winter Olympics: (**a**) If we plot the first racer's horizontal position on the track over time, we get a sine curve with a particular frequency. If the second racer circles the inner track with a frequency that is significantly higher than the first racer's frequency, the racers pass each other at the exchange point, and the second racer receives only a very weak push. (**b**) However, when the two racers have similar frequencies, they are at the same location on the track at the exchange point, and the second racer receives a strong push

the Fourier Transform tests our NMR data with different sine curves. If the second racer's trips around the track are more frequent than the previous racer's, then she may pass her teammate in-transit and miss out on an efficient push (Fig. 1.9a). When her trips around the track are less frequent than the first racer, then her teammate will be on the opposite side of the track when she enters the race, and, thus again, the second racer will miss out on the speed boost. However, when the second racer's frequency perfectly matches her teammates, every trip around the track will result in an efficient push (Fig. 1.9b). In other words, contact time is maximum when the frequencies of the two racers are exactly in sync.

Okay, this frequency matching strategy is a bit silly because the optimal speed for each racer is a obvious in this situation. However, the analogy mirrors beautifully how the Fourier Transform finds the frequencies in our NMR data: It tests every possible sine curve, and when the frequencies are near the frequency of the sine curve (or curves) in the raw time-domain NMR data (Fig. 1.7a), then the Fourier Transform returns large positive values or intensities (Fig. 1.7b).

Mathematical Sidebar 1.1: Fourier Transform

Mathematically, the Fourier Transform (FT) multiplies the Crosby, Stills, and Nash chord at each time point by a sine function and then sums up all these products. Let's simplify the chord down to one note with a frequency of 500 MHz (Fig. 1.10a). When the FT multiplies this note by a sine function with a frequency close to 500 MHz (for example, 499 MHz), the product will have a large positive value because the two oscillating curves tend to have positive values at the same time and negative values

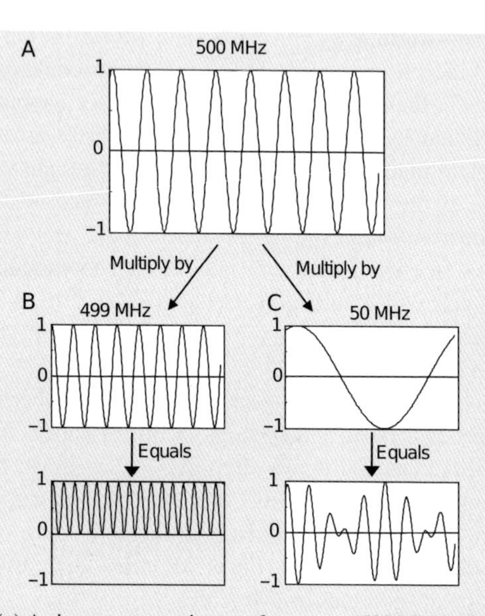

Fig. 1.10 (**a**) A sine wave contains one frequency (500 MHz). (**b**) When the curve is multiplied by another sine curve of a similar frequency (499 MHz), every point on the resulting curve is positive (>0). (**b**) In contrast, when this curve is multiplied by a sine function with an extremely different frequency (50 MHz), the resulting curve has equal positive and negative values

at the same time (Fig. 1.10b), and (+) times (+) or (–) times (–) gives a positive value (Fig. 1.10b). In contrast, if the Fourier Transform multiplies the note by a sine curve with a frequency far away from the note (say 10 MHz), the product will be approximately zero because when the sine function is positive, the chord will be positive 50% of the time and negative the other 50% of the time (Fig. 1.10c). Therefore, the Fourier Transform has a large positive value at the note's frequency (Fig. 1.11d) and a value of almost zero at every other frequency. The mathematics of this process is quite elegant and given by the equation:

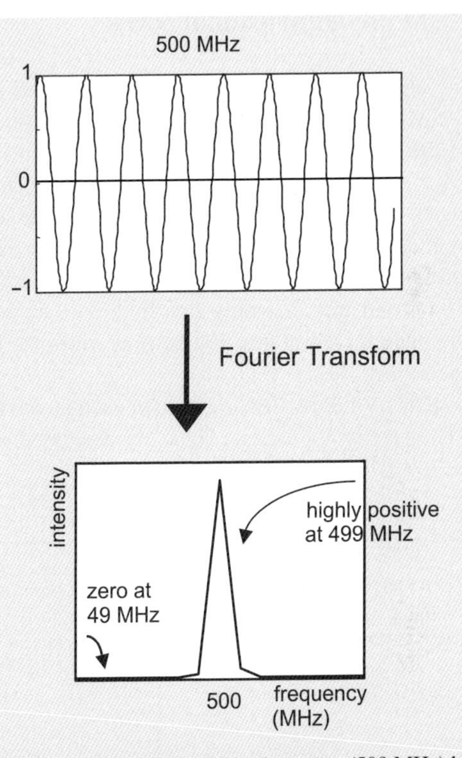

Fig. 1.11 The Fourier Transform of this sine curve (500 MHz) has a maximum at 500 MHz, a large positive value at 499 MHz, and a value of zero at frequencies far away from 500 MHz

$$I_{\text{spectrum}}(f) = \int_{0}^{\infty} I_{\text{FID}}(t)\sin(2\pi f t)dt$$

Where I_{spectrum} is the intensity of the spectrum at frequency f and I_{FID} is the intensity of the NMR signal (i.e., the free induction decay or FID) at time t.

1.5 Basics of One-Dimensional NMR

Now we have all the concepts required to understand and use one-dimensional NMR spectra, like the one shown in Fig. 1.12 for the hydrogens of the amino acid threonine (Fig. 1.12, inset). In the spectrum, there is a "peak" at each frequency present in the NMR data, corresponding to each hydrogen ringing in the molecule. The *y*-axis tells us "how much" of that frequency is present. For example, there are three hydrogens at the γ position of threonine (all in the same electromagnetic environment), but only one hydrogen at the β position. Therefore, the peak for the γ position is three times the height of the β-hydrogen peak (Fig. 1.12).

The *x*-axis tells us the frequency at which each atom rings, but look at the units. They are in "ppm," not Hz as we discussed earlier. "ppm"

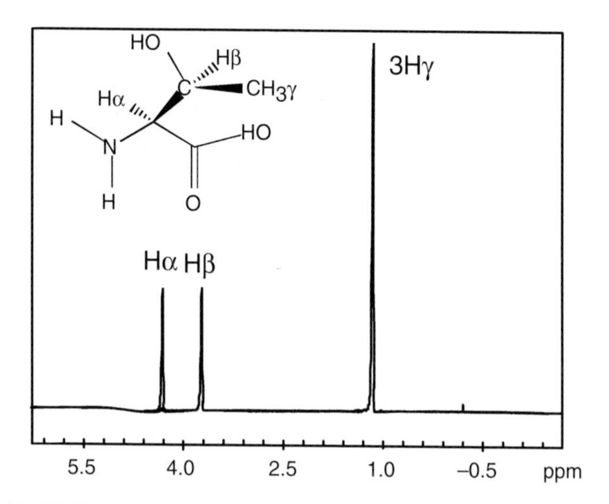

Fig. 1.12 NMR spectrum for the amino acid threonine, showing peaks for the Hα, Hβ, and Hγ atoms

is short for "parts-per-million," and it is very similar to the concept of "percent" (or 1/100). For example, let's say you want to loose weight. You start off at 130 lbs, and after dieting for 2 months, you now weigh 125 lbs. What percentage weight did you loose? That's easy, right? You merely subtract 130–125 and then divide by 130, which is 0.04 or 4% (or 4 parts-per-hundred). You calculate "ppm" in NMR the same way:

You simply subtract the atom's ringing frequency from a reference frequency and then divide by the reference frequency (see Mathematical Sidebar 1.2). In the diet example, your initial weight of 130 lbs is the point of reference. For NMR, the reference is usually the frequency of a compound called trimethylsilane (or TMS). The resulting value is tiny, 0.000010 or less. So instead of writing all those zeros, we merely multiply by one million (10^6 or 1,000,000) and slap "parts-per-million" (ppm) after the number. It is just like multiplying by 100 and using % after a value (i.e., 0.04 = 4%).

Using "ppm" values greatly simplifies the *x*-axis. Instead of having to write 500.00212 MHz for the frequency at which the β-hydrogen rings in an 11.75-T magnet, we can just say 4.24 ppm. Similarly for the γ-proton, instead of 500.000658 MHz, we say 1.135 ppm. You can definitely see the advantage of ppm over Hz. This "ppm" value is called the atom's *chemical shift*, and it's a term you will hear often in NMR. All you need to remember is that chemical shift is how fast the atom rings relative to some reference frequency.

Another advantage to using chemical shift is that it stays constant across different magnetic field strengths, unlike the atom's ringing frequency (remember, the ringing frequency increases as the strength of the surrounding magnetic field increases). For example, the frequency for the β-hydrogen at 18.8 T is 800.003392 but only 500.00212 at 11.75 T. In both fields, the chemical shift is 4.24 ppm. This allows NMR spectroscopists to compare easily experiments and spectra run at different field strengths.

Mathematical Sidebar 1.2: Converting Hz to ppm

The chemical shift in parts-per-million (ppm) is denoted by δ and is calculated using the following equation:

$$\delta = \frac{(\text{resonance frequency of the atom-resonance frequency of TMS})*10^6}{\text{operating frequency of the magnet}}$$

where TMS is tetramethylsilane, a compound with a ringing frequency that usually does not overlap with that of atoms in proteins and other biological molecules.

For example, for the water proton at 500 MHz:

$$\delta = \frac{(500.00212\ \text{MHz} - 500.00000\ \text{MHz})*10^6}{500\ \text{MHz}} = 4.24\ \text{ppm}$$

Further Reading

Bloch F (1953) The principles of nuclear induction. Science 118:425–430.

Levitt M (2001) *Spin dynamics: basics of nuclear magnetic resonance*, Chapters 1–3 and 5. Wiley, Chichester.

Wüthrich K (1986) *NMR of proteins and nucleic acids*, Chapters 1–3. Wiley, Chichester.

Cavanagh J, Fairbrother WJ, Palmer AG III, Rance M, Skeleton NJ (2007) *Protein NMR spectroscopy: principles and practice*, 2nd edn., Chapter 1. Academic, Amerstdam.

Cantor CR, Schimmel PR (1980) *Biophysical chemistry part II: techniques for the study of biological structure and function*, Chapter 9. W. H. Freeman and Company, New York.

Chapter 2
Bonded Bells and Two-Dimensional Spectra

Falling dominos. A bat hitting a baseball out of the stadium. A dog tugging on a leash. Energy transfer is all around us (Fig. 2.1). One of the most elegant demonstrations of energy transfer is "Newton's cradle" (Fig. 2.2). You've probably seen this executive toy: five metal balls are attached to a metal frame by a thin wire, like five pendulums kissing each other. When the first ball is swung (Fig. 2.2a), it hits the neighboring balls, but only the ball at the extreme end reacts and swings out (Fig. 2.2b). The kinetic energy of the first ball is almost perfectly transmitted through the middle balls, causing only the ball at the end to move.

Like the metal balls in "Newton's cradle," ringing atoms in molecules also transfer their excess energy to other atoms. NMR spectroscopists call this energy transfer *coupling*. This phenomenon is very useful for characterizing the details of molecular structures because it tells us which atoms are close to one another. We'll see how this works in the next two chapters.

2.1 Introduction to Coupling

In Chap. 1, we learned that atoms ring like bells when they are in a strong magnetic field. The frequency at which they ring depends on the atom type and its electromagnetic environment (i.e., the types of atoms

M. Doucleff et al., *Pocket Guide to Biomolecular NMR*,

DOI 10.1007/978-3-642-16251-0_2,

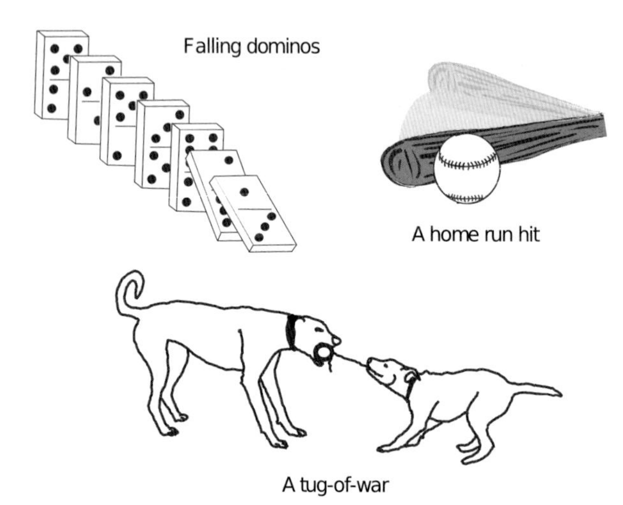

Fig. 2.1 Three common examples of energy transfer: dominos knocking each other down, a bat hitting a baseball, and two dogs playing tug of war with a toy. Can you identify more examples of energy transfer around you right now?

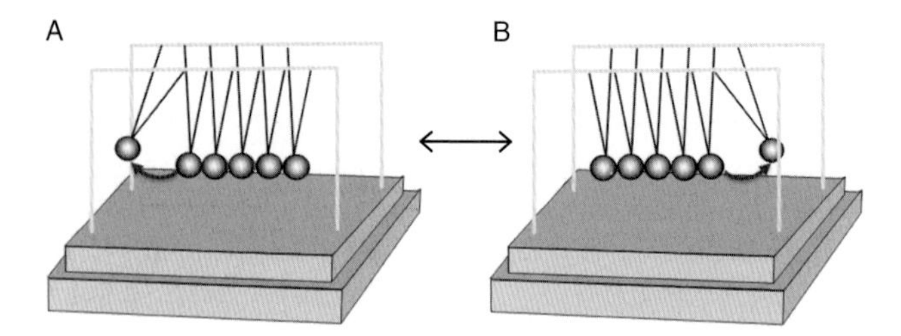

Fig. 2.2 "Newton's cradle": As the two metal balls on the outside swing back-and-forth (**a** and **b**), the middle balls transfer the kinetic energy almost perfectly and thus stand completely still

around it). In an NMR experiment, we make atoms ring by applying a strong radiofrequency pulse to the molecule. This pulse hits particular atoms like a big hammer, causing the nuclei to ring and create the NMR signal (Fig. 1.6).

Like a church bell ringing, a ringing atom has excess energy, and it can transfer this surplus energy to neighboring atoms if the two atoms are:

(a) Connected by a covalent bond

or

(b) Close in space, more specifically less than about 5 Å (or 5×10^{-10} m) apart.

In both cases, the type of energy exchanged is nuclear magnetization. You'll hear and read this term often in NMR. For example, "The magnetization is transferred from atom A to atom B." You can think of it like the metal balls in Newton's cradle (Fig. 2.2), moving energy from one nucleus to another. In Newton's cradle, the extra energy is kinetic; in NMR, the extra energy is electromagnetic.

In the language of physical chemistry, transfer of magnetization through bonds is referred to as "J-coupling," while transfer of magnetization through space is known as "dipole-dipole" coupling. Before we take a look at J-coupling (we'll discuss dipole-dipole coupling in the next chapter), let's examine the term *coupling* a bit further.

In the non-NMR world of everyday life, when you hear the word *couple*, you probably think of two people dating or a married couple. Indeed, when two people are a couple, their activities are connected. They eat together, work-out together, and study together. After years of marriage, couples even start dressing the same—a "style coupling"—or eating similar foods—a "diet coupling."

In NMR, *coupling* is quite similar to the more common, everyday usage of the word. When two nuclei in a molecule are coupled, the activity of one nucleus affects or controls the activity of the other nucleus. When one nucleus starts ringing, it makes the neighboring nucleus join in. Think back to Newton's cradle (Fig. 2.2). When the ball on the right swings, it causes the metal ball on the left to swing as well. Thus, the two balls are coupled. The first ball shared its kinetic energy with the second ball. In NMR, coupled nuclei also share each other's excess energy, just like human couples share so many things in life.

What about when a married or dating couple starts doing the opposite actions of each other. For example, you are so sick of your boyfriend that when he goes to the library to study, you intentionally stay in the dorm and work. You are still *coupled* because his choices strongly affect your choices. However, instead of your actions being *correlated*, they are now *anti-correlated.*

2.2 Bonded Bells: J-Coupling

Two atoms connected by a covalent bond can share or transfer their energy when they ring. This is referred to as J-coupling, and it is extraordinarily useful in biomolecular NMR. It tells us not only which atoms are bonded to each other but can also provide information on angles between neighboring bonds (known as dihedral angles). Let's learn how it works.

Two atoms are covalently bonded when their nuclei share electrons. These electrons orbit between both nuclei, keeping the atoms close together (Fig. 2.3). You can think of these shared electrons as springs connecting the nuclei (Fig. 2.4). If we start ringing one nucleus, say the nitrogen atom (Fig. 2.4a), the spring will oscillate and eventually transfer part of the energy to the attached hydrogen, causing it to ring too (Fig. 2.4b). This is the essence of J-coupling.

Although the analogy to a spring is an oversimplification, it does help explain a few important and useful characteristics of the J-coupling phenomenon:

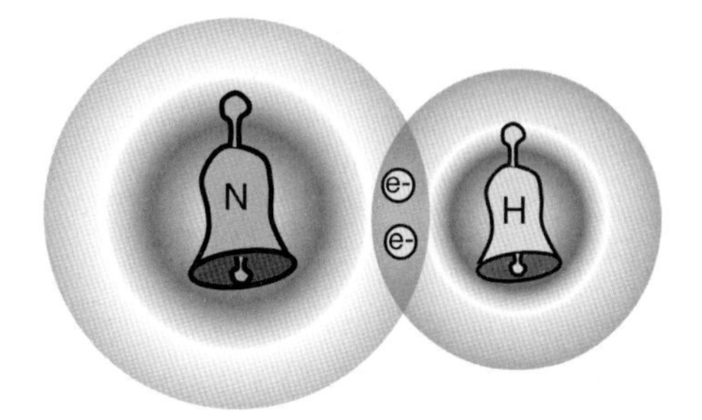

Fig. 2.3 Two atoms are covalently bonded when they share two or more electrons

Fig. 2.4 (**a**) When we ring the nitrogen's nucleus, (**b**) the extra energy can transfer through the shared electrons and ring the neighboring hydrogen. Thus, the shared electrons in a covalent bond act like a spring between the nuclei

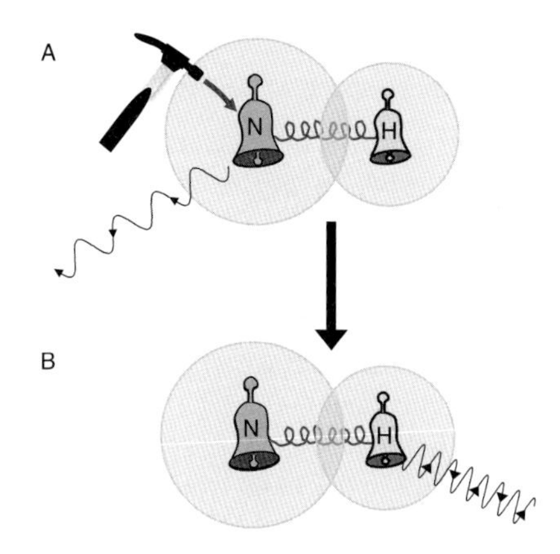

(a) As a spring oscillates back-and-forth between two endpoints, energy (or magnetization) oscillates between two bonded nuclei.
(b) How fast a spring oscillates depends on the material and other properties of the spring. For atoms, the speed of the energy exchange depends on the particular properties of the electrons and the bond between the two atoms.

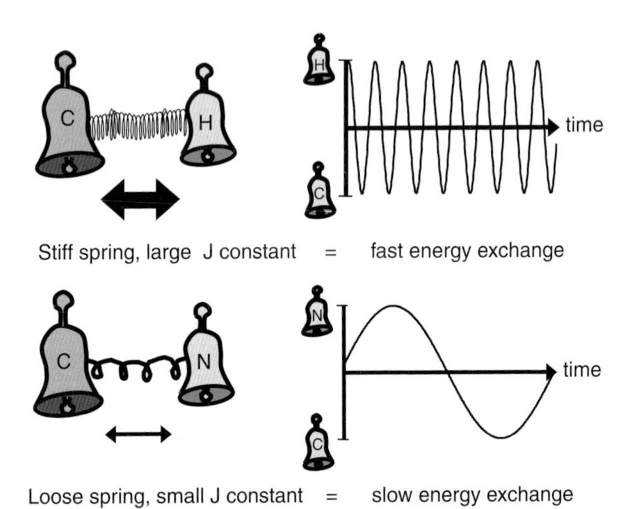

Fig. 2.5 (**a**) Bonds with large J-coupling constants (stiff springs) transfer excess energy between the two nuclei quickly and (**b**) bonds with small J-coupling constants (loose springs) transfer the energy more slowly

Stiff springs, like the ones in car shocks, have large "spring constants" (Fig. 2.5a). In this case, when you hit the nitrogen, the spring will oscillate quickly, and the energy moves back-and-forth between the atoms at a high frequency. Loose springs, like a slinky toy, have small "spring constants." In this case, when you hit the nitrogen atom, the spring will oscillate slowly (Fig. 2.5b), and the energy will move back-and-forth between the two nuclei at a much slower frequency. The "NMR spring constant" for bonded nuclei is called the *J-coupling constant*, and it tells us how fast the shared electrons transfer energy (or magnetization) back-and-forth between two bonded nuclei.

Let's look again at the valine–threonine peptide as an example. In Fig. 2.6, the values next to each bond show the bond's J-coupling constant. Notice that the units are frequency—Hertz. Why? Because the J-coupling constant (also known as the scalar coupling constant) tells us how fast energy oscillates between the two connected atoms, just

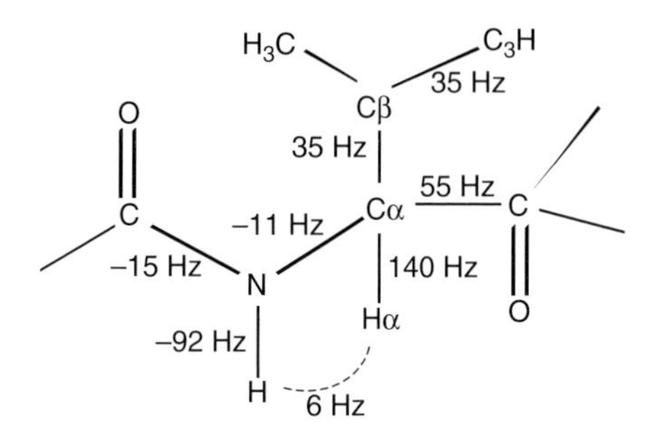

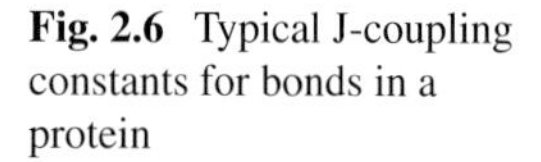
Fig. 2.6 Typical J-coupling constants for bonds in a protein

like the chemical shift from Chap. 1 tells us how fast a nucleus rings. It's almost like the bond is ringing!

Bonds with large J-coupling constants, such as the one connecting ^{13}Cα and ^{1}Hα atoms ($^{1}J_{C\alpha H\alpha}$ ~ 140 Hz), will quickly transfer magnetization between the two bonded atoms (Fig. 2.5a). Thus, the energy state of the ^{1}Hα nucleus greatly affects the state of the ^{13}Cα nucleus. If the ^{1}Hα nucleus starts ringing strongly, the ^{13}Cα nucleus will definitely join in and start ringing too. This is analogous to how your mood has a profound effect on your spouse's mood when you are "tightly coupled."

In contrast, bonds with small J-coupling constants, such as the one connecting ^{13}Cα and ^{15}N atoms ($^{1}J_{C\alpha N}$ ~ 11 Hz), take more time to transfer magnetization (Fig. 2.5b). These atoms are more "weakly coupled," and their energy states barely affect each other. If one nucleus is ringing strongly, the other nucleus is hardly affected. Perhaps you are "weakly coupled" to your parents right now—if they nag you long enough, you might do what they say, but it takes time and effort.

Notice that $^{1}H_{N}$ and ^{1}Hα atoms in Fig. 2.6 also have a J-coupling constant (6 Hz). Although they are not directly bonded, they are connected via three bonds: H_N to N, N to Cα, and final Cα to Hα. This is called a *three-bond coupling* (denoted by a ^{3}J coupling constant). Although ^{3}J coupling constants are usually quite small, as it is here—only 6 Hz, they can provide useful structural information

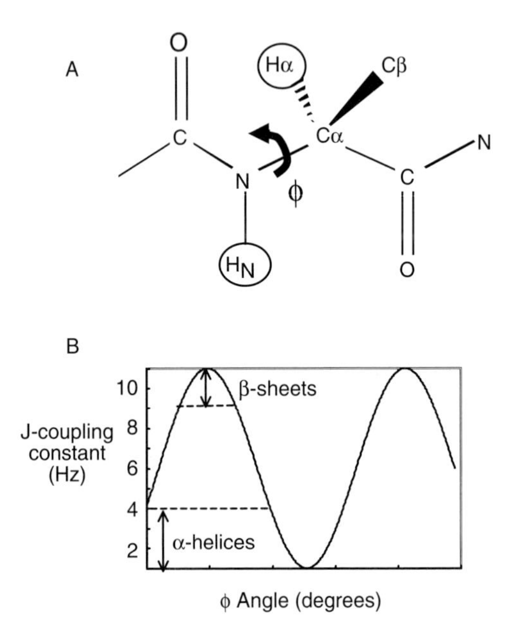

Fig. 2.7 The J-coupling constant between the H_N and the H_α atoms in a protein tells us (**a**) the approximate angle between these two atoms (the torsion angle φ about the N–C_α bond) and (**b**) whether this residue of the protein forms an α-helical or β-sheet configuration

for macromolecules. For example, the ^1H_N–$^1H\alpha$ ^3J-coupling constant tells us the approximate value of the ϕ (phi) torsion angle about the N–Cα bond (Fig. 2.7a). In α-helices of proteins, ϕ is approximately −60°, and the $^3J_{HN\alpha}$ coupling constant is barely detectable at about 4 Hz (Fig. 2.7b). In contrast, extended regions of the protein (called β-sheets) have ϕ values around −135°, and the $^3J_{HN\alpha}$ is larger (~9 Hz) (Fig. 2.7b).

The relationship between ϕ and the $^3J_{HN\alpha}$ coupling constant is described by a geometric relationship known as the Karplus relationship (see Mathematical Sidebar 2.1). It is a great place to start when determining the structure of a protein because with one easy measurement, we find where the protein forms α-helices or β-sheets. Values less than 4 Hz identify helical regions and those greater than 9 Hz most likely form extended or β-sheets regions (Fig. 2.7b).

Before moving on, we want to emphasize one key point about J-coupling: It involves energy or magnetization transfer between two nuclei via the electrons in a covalent bond(s). Therefore, J-coupling occurs *only* between nuclei that are connected by a small number of covalent bonds. If the connection involves more than three bonds, the J-coupling constant is negligible, and the atoms are, for all practical purposes, not J-coupled.

Mathematical Sidebar 2.1: Karplus Equation

The Karplus equation describes the relationship between the dihedral angle and the vicinal coupling constant. This was first applied to the configurations of ethane derivatives but has been expanded for use in a number of systems, including peptides. The general form of the equation is

$$J(\phi) = A\cos^2\phi + B\cos\phi + C$$

where J is the three-bond coupling constant between two atoms; ϕ is the dihedral angle involving these two atoms and the two atoms connecting them by bonds; and A, B, and C are empirically derived parameters that depend on the atoms. In proteins, J couplings are largest when the torsion angles are $-135°$ (β-sheet) and smallest when the angle is $-60°$ (α-helix).

2.3 NMR Maps: Two-Dimensional Spectra

Okay, when two atoms are connected by a covalent bond, energy (or more specifically magnetization) transfers between the two nuclei via the bonded electrons. What good is that? For starters, we use this phenomenon to begin matching up specific atoms in the molecule with the ringing frequencies found in the spectra. Let's see how.

To characterize molecular structures and dynamics by NMR, the first step is to determine the ringing frequency (or chemical shift) for each atom in the protein. This is called *chemical shift assignments*, and it can be quite tricky. Think of a "chicken-and-the-egg" problem combined with a huge "connect-the-dots," or a Suduko puzzle with 10,000 squares!

If you are interested only in the biophysical properties of the backbone atoms of a protein, then you need only to determine the chemical shifts for the backbone atoms. But if you want an NMR image of the entire protein, you need to assign the ringing frequency for almost every atom in the protein, which is easily over a 1000 atoms.

For chemical shift assignments of proteins, NMR spectroscopists almost invariably start with the backbone amide protons ($^{1}H_{N}$)—the hydrogens attached to the nitrogens in the peptide bond (Fig. 2.6). Figure 2.8a shows the NMR spectrum for the amide protons in calmodulin, a 148-residue protein found in all eukaryotic organisms. Notice the limits of the *x*-axis: 5.5–10.5 ppm. This is the frequency range for virtually all amide protons in all proteins. In general, each residue of calmodulin (with the exception of the N-terminal residue and prolines) has one amide proton, so there is approximately one peak for each residue of the protein. For calmodulin, this means we have ~148 peaks between 5.5 and 10.5 ppm.

With so many peaks in such little space, it's tough to pick out specific hydrogens and say, "this peak must belong to the amide proton of residue 29" (arrow in Fig. 2.8a). To do this, we must spread the peaks out. In other words, we need to *increase the resolution* of the spectrum. How could we do this?

1. For starters, we could make the peaks narrower (i.e., sharper). Unfortunately, broad peaks are inherent in protein NMR, and, as we'll learn in Chaps. 4 and 5, they don't slim down easily.
2. Another option is to run the experiment at a stronger magnetic field—say, for example, 900 MHz versus 500 MHz. Why would that

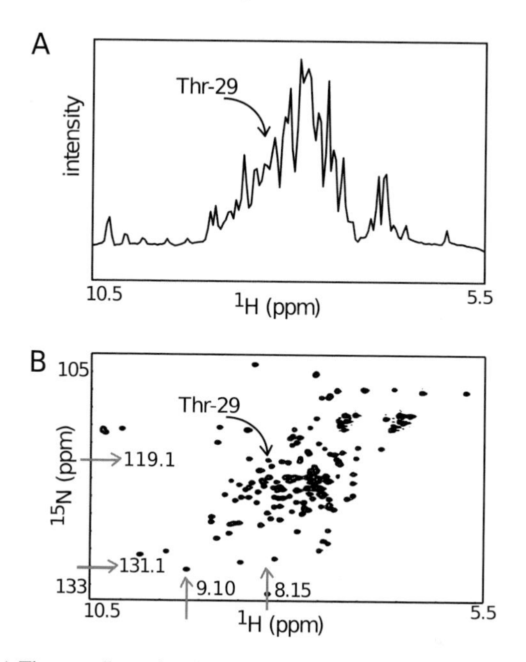

Fig. 2.8 (**a**) The one-dimensional NMR spectrum of a protein's amide hydrogens (H_N). (**b**) The two-dimensional version of the same spectrum (an ^1H–^{15}N HSQC spectrum)

help spread the peaks out? (See the end of Chap. 2 for the answer.) Unfortunately, higher field magnets aren't usually available.

3. In the end, the trick that works best is to *add another dimension*. Why does this increase our resolution?

Think about driving to the mountains. When you are far away on level terrain, all the peaks flatten into one plane like a vertical pancake (Fig. 2.9a). You know that your favorite peak for skiing is straight ahead, but it's difficult to identify the exact one. The problem is that you can see vertically, but not horizontally. Now think about looking at these mountains from an airplane far above: You can easily identify

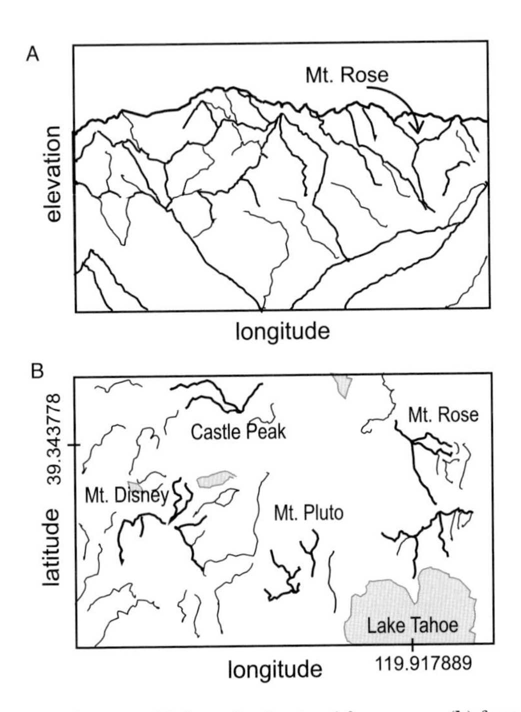

Fig. 2.9 A mountain range (**a**) from the front and far away or (**b**) from above

individual peaks because you see in all directions—east to west and north to south (Fig. 2.9b). Adding the extra dimension separates and spreads the peaks out.

In NMR, we add extra dimensions by transferring magnetization between coupled atoms, such as J-coupled nuclei. This is called "two-dimensional" NMR, and it involves four basic steps[1] (Fig.2.10):

---

[1]This is an oversimplification of the experiment, but it provides the essence of two-dimensional NMR. We'll describe a bit more of the details below, but if you're still not satisfied, check out Chap. 7 of "Protein NMR Spectroscopy: Principles and Practice" by Cavanagh et al. (2007).

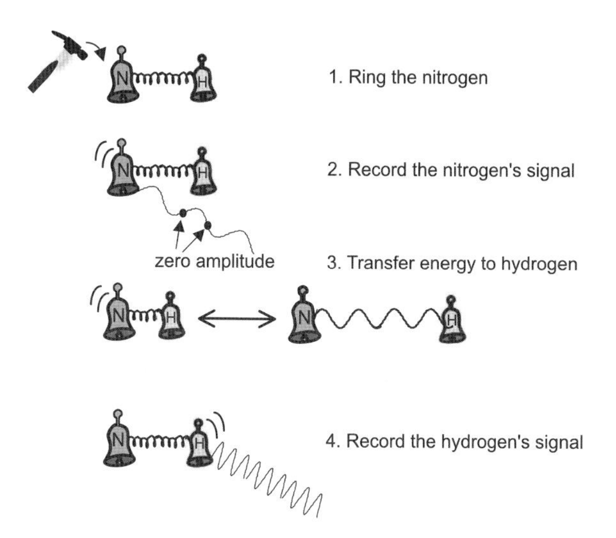

Fig. 2.10 The four basic steps of the ^{1}H–^{15}N HSQC experiment for connecting the chemical shift of an amide hydrogen with the chemical shift of its covalently attached nitrogen

1. Ring the nitrogen atom.
2. Record its electromagnetic signal.
3. Let the nitrogen ring the attached hydrogen via J-coupling.
4. Record the electromagnetic signal of the hydrogen.

The result is a two-dimensional map of the NMR spectrum (Fig. 2.8b), similar to the aerial view of the mountains (Fig. 2.9). The ringing frequencies (or chemical shifts) of the amide hydrogens are along the x-axis and the ringing frequencies for the attached nitrogens are along the y-axis (Fig. 2.8b). With the extra elbow-room of an additional dimension, many peaks are now completely separated and distinct, making the spectrum easy to analyze and significantly more useful (compare Fig. 2.8a to 2.8b).

In this spectrum, there is a peak for each $^{1}H_N$ atom bonded to a ^{15}N atom, or each "$^{1}H_N$–^{15}N pair." (To learn why we study ^{15}N atoms with NMR instead of the more common ^{14}N atoms, see Mathematical Sidebar 2.2.) The location of the peak is at the point $x =$ hydrogen chemical shift, $y =$ nitrogen chemical shift. You can read this two-dimensional spectrum in the same manner that you read a map on a longitude–latitude grid. Say your favorite place to ski is Mount Rose. How would you find Mount Rose on a map if you know it is located at 39.343778° latitude (x-axis) and 119.917889° longitude (y-axis)? In Fig. 2.9b, you could draw a horizontal line passing through the y-axis at 119.917889° and a vertical line passing through the x-axis at 39.343778°. Mount Rose is where the two lines intersect.

Analogously, you can find the cross-peak for threonine-29 on the two-dimensional NMR map in Fig. 2.8b if you know the chemical shifts (or ringing frequencies) for its amide hydrogen and nitrogen. Say the amide hydrogen for threonine rings at 8.15 ppm, and the amide nitrogen rings at 119.1. To find the cross-peak of threonine-29 in the two-dimensional spectrum, simply draw a vertical line through 8.15 ppm on the x-axis and a horizontal line through 119.1 ppm on the y-axis—voila! Like Mount Rose, threonine-29 is where the two lines intersect. Since this peak is now distinct from all the others, we can use it as a probe to characterize the structure and dynamics of residue 29 in calmodulin. We'll see how this works in later chapters.

You can probably see now how this two-dimensional spectrum (Fig. 2.8b) helps us start matching up atoms with their specific ringing frequencies. Say you know that residue 137's amide proton rings at 9.10 ppm, but you don't know the chemical shift for its attached nitrogen. To make this chemical shift assignment, simply draw a vertical line through $x = 9.10$ ppm in the two-dimensional spectrum. This line cuts straight through a peak at $y = 131.1$, telling us that the amide nitrogen rings with a frequency of 131.1 ppm.

Obviously this procedure wouldn't work in the center of the spectrum where multiple peaks overlap and pile on top of each other (Fig. 2.9b). In this case, we would need to add another dimension to spread these peaks out further. We'll talk more about these "three-dimensional" spectra in the next chapter, but for now, let's focus on the most important spectrum in biomolecular NMR: the HSQC.

Mathematical Sidebar 2.2: Why ^{12}C and ^{14}N Atoms Are So Shy?

When studying a protein by NMR, one of the first steps is to create a "^{15}N-labeled protein" in which all the ^{14}N atoms are replaced with ^{15}N atoms. Then to solve the structure of the protein NMR, we also need to replace the ^{12}C atoms in the protein with ^{13}C atoms to create "^{13}C and ^{15}N-labeled protein." But why? Why can't we study ^{12}C and ^{14}N versions of proteins? To answer this question, we need to learn more about why atoms ring in the first place.

Atoms carry a few intrinsic properties that dictate how they respond to forces in nature. First, atoms possess mass. The more mass an atom carries, the more it responds to gravitational fields. Atoms also have a charge: positive, negative, or neutral. The charge on the atom tells us how the atom responds in an electric field. Although we don't usually "see" the charge, we can observe it (e.g., lightning bolts). Sometimes we can even feel the charge of atoms. Simply rub your feet across a carpet and then touch a friend, and you'll definitely "feel" charge.

The third property of atoms is *spin*. The spin tells us how an atom responds to a magnetic field. Like charge, spin comes in multiple flavors:

- Nuclei with even mass numbers (i.e., the total number of protons and neutrons) and even numbers of protons have spins of zero. For example, ^{12}C has a spin of zero.
- Nuclei with even mass numbers and odd number of protons have integral spin numbers. For example, ^{14}N has a spin of one.
- Nuclei with odd mass numbers have half-integral spins: 1/2, 3/2, 5/2, etc. For example, ^{1}H, ^{13}C, or ^{15}N each have a spin of 1/2.

Each flavor of spin acts differently in a magnetic field. Those with a spin of zero, like ^{12}C, are very shy and don't ring at all in the magnet. Thus, all those ^{12}C atoms in proteins are not very useful for structure determination by NMR. Nuclei with spins greater than ½ (i.e., spins of 1 and larger) do ring in the magnetic field, but, unfortunately, their ringing is very short. These atoms possess what is called an "electric quadrupolar moment," which dampens their ringing so much that detecting their signals becomes extremely difficult. Thus, ^{14}N atoms in proteins are also not useful for NMR studies.

Nuclei with spins of 1/2 are perfect! They ring in a magnetic field, and they don't possess "electric quadrupolar moment." So their signals stay strong and loud for a long enough time that we can easily detect them. Thus, for biomolecular NMR spectroscopy, the most important nuclei are those with spins equal to 1/2: ^{1}H, ^{13}C, and ^{15}N. *Can you name another nuclei often found in biological molecules that could have a spin of 1/2?*

2.4 The ^{1}H–^{15}N HSQC: Our Bread and Butter

The spectrum we've been analyzing in Fig. 2.8b is called a ^{1}H–^{15}N *H*eteronuclear *S*ingle-*Q*uantum *C*oherence correlation spectrum, or HSQC for short, and we create this spectrum using the steps outlined in Fig. 2.10. You need to understand this spectrum well because it's the starting point for almost all other biomolecular NMR experiments. In other words, it's our "bread and butter" experiment.

The ^{1}H–^{15}N HSQC spectrum shows us which protons and nitrogen atoms are connected by a single covalent bond. There are also ^{1}H–^{13}C HSQC spectra that tell us what protons are attached to what carbons, but we'll focus on the ^{1}H–^{15}N correlation experiment since it is the spectrum most often seen in protein NMR publications. (To learn why we study ^{13}C atoms with NMR instead of the more common ^{12}C atoms, see Mathematical Sidebar 2.2.)

Here are the five most important attributes of the ^1H–^{15}N HSQC for proteins. Read them, understand them, remember them:

(a) **The Axes**: The *x*-axis gives the ringing frequency or chemical shift of each protons attached to a nitrogen atom. The *y*-axis gives the ringing frequency of each nitrogen atom attached to a proton.
(b) **The Cross-peaks**: A cross-peak appears in the center of the spectrum wherever a hydrogen is bonded to the nitrogen.
(c) **Total Number of Peaks**: The total number of peaks is approximately equal to the total number of residues in the protein:[2]

- If the spectrum has more peaks than this, the protein is probably adopting multiple conformations or forming heterogenous oligomers (i.e., heterodimer, heterotrimer, etc.).
- If the spectrum has less peaks than predicted, regions of the protein are probably dynamic, exist in multiple confirmations, or partially unfolded.

(d) **Peak Intensity and Shape**: For "well-behaved" proteins, all peaks should have approximately the same linewidth and therefore peak height, like the spectrum in Fig. 2.11a. If a significant number of peaks are weak and broad (or completely missing), regions of the protein are probably dynamic and/or exist in multiple conformations.
(e) **The Peak Pattern**: Peaks should be relatively well separated with minimal overlap (Fig. 2.11a); this is a good sign that the protein is well folded and not aggregating. Otherwise

[2]Specifically, the number of peaks in a ^1H–^{15}N HSQC spectrum of a protein is given by the number of residues plus two times the number of glutamines and asparagines minus the number of prolines minus one (for the N-terminal residue, which has a rapidly exchanging NH_3 group instead of an NH group).

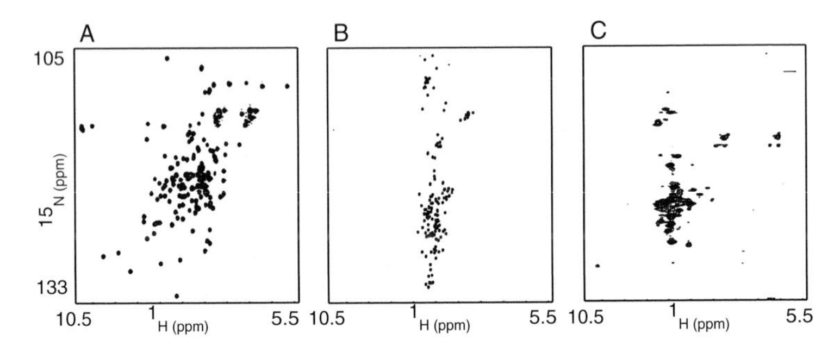

Fig. 2.11 Examples of ^{1}H–^{15}N HSQC spectra for (**a**) a well-behaved protein folded into one configuration, (**b**) a completely denatured, unfolded protein, and (**c**) an aggregated protein in multiple configurations

- if peaks are sharp but largely grouped together between $x = 7.5$ and 8.5 ppm (Fig. 2.11b), the protein is probably unfolded in a random coil-like configuration—think of a string of spaghetti in boiling water;
- if the peaks coalescence into one big blob near the center of the spectrum, the protein is probably aggregating or forming high-order oligomers (Fig. 2.11c).

As this list demonstrates, the ^{1}H–^{15}N HSQC provides substantial information about the state of the protein even before we assign the chemical shifts to specific atoms. In fact, crystallographers and other biophysicists often use the ^{1}H–^{15}N HSQC to check if their proteins are folded and well-behaved before they set up crystallization screens or perform other time-consuming experiments.

Before we go on, test your understanding of the ^{1}H–^{15}N HSQC spectrum with the following question:

The sidechains of glutamine and asparagine residues have amide groups with two hydrogens attached to one nitrogen. With this knowledge, which peaks in Fig. 2.8b belong to glutamine and asparagine sidechains? Explain why you have selected these peaks.

2.5 Hidden Notes: Creating Two-Dimensional Spectra

For the last part of Chap. 2, let's look closer at how two-dimensional spectra, such as the ^{1}H–^{15}N HSQC in Fig. 2.8a, are created. We'll use an analogy to clarify the confusing world of "indirect" dimensions.

Do you ever listen to talk radio or sport broadcasts on AM radio? Although its super old school, AM radio has an endearing quality that takes you back in time. Its characteristic raspy, tinny sound is a direct result of AM's broadcasting style, which shares surprising similarities with two-dimensional NMR.

How does AM radio work? In a nutshell, the AM radio in your car converts an electromagnetic wave into sound waves through your speakers (Fig. 2.12). Each AM radio station in an area is assigned a specific frequency at which to broadcast its signal. For example, in Washington, DC, WCYB, which plays gospel music, broadcasts its signal on an electromagnetic wave with a frequency of 1340 kHz (Fig. 2.13b). On the other hand, WTEM, which plays local news and talk shows, broadcasts its signal on an electromagnetic wave with a frequency of 630 kHz (Fig. 2.13a). When you dial in "AM630" on your radio, you're telling the radio "pick up only waves with a frequency of 630 kHz and ignore all other frequencies."

For simplicity, let's say that both AM630 and AM1340 are broadcasting one note each (remember, a note is merely a sound wave at a particular frequency): AM630 is playing the note

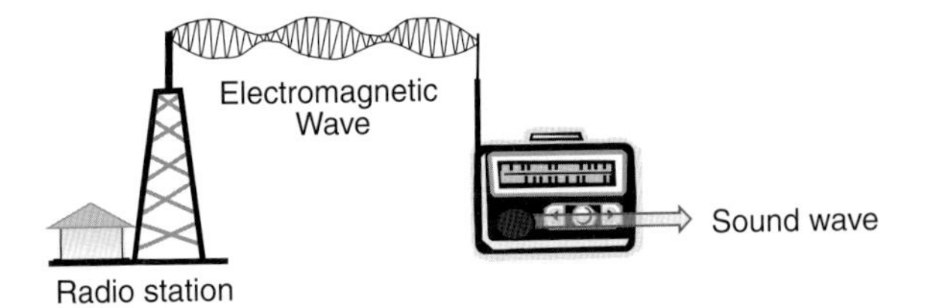

Fig. 2.12 AM radio stations broadcast a signal with oscillating amplitude. Your radio converts it into a sound wave by analyzing the amplitude changes in the electromagnetic wave

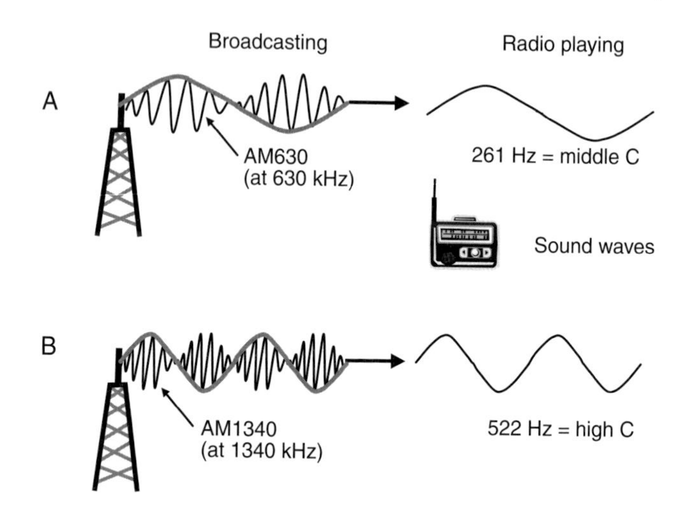

Fig. 2.13 (**a**) AM630 broadcasts its electromagnetic signal at 630 kHz, but the wave's amplitude oscillates at 261 Hz, the frequency of the note middle-C. (**b**) AM1340 broadcasts its signal at 1340 kHz, but the wave's amplitude oscillates at 522 Hz, which the radio plays high-C

"middle-C" = 261 Hz (Fig. 2.13a), while AM1340 is playing a C note one octave higher = 522 Hz (Fig. 2.13b). How does each station tell your car radio to play the right note? The radio station can't actually send out a wave at the note's frequency because the station can broadcast only at its assigned frequency. So the radio station needs to hide or imbed the note's frequency inside its electromagnetic wave. AM radio accomplishes this by varying the wave's amplitude (i.e., its maximum height) at the frequency of the specific note (Fig. 2.8a). Thus, your radio knows what sound wave to produce in your speakers by following the fluctuations in the electromagnetic wave's amplitude. This type of broadcasting is called "Amplitude Modulation," or "AM" because the signal is created by modulating the electromagnetic wave's amplitude.

The note on AM1340 is twice the frequency of AM630's note, so AM1340 changes its amplitude twice as fast as AM630 (Fig. 2.13a versus 2.13b). You can easily see these hidden notes in Fig. 2.13 by drawing a line along the edge or amplitude of each station's radio wave (gray lines). The resulting sine curves have the exact frequencies of the sound waves played by the radio stations. If we plot the radio stations' frequencies, 630 and 1340 kHz, on the *x*-axis and the frequency of the note they broadcast on the *y*-axis, 261 and 522 Hz, respectively, the resulting graph looks quite similar to the two-dimensional ^{1}H–^{15}N HSQC spectra in Fig. 2.11a. This is no coincidence! Like AM broadcasting, the peaks in a two-dimensional NMR spectrum are also created by modulating the amplitude of the NMR signal.

For the ^{1}H–^{15}N HSQC in Fig. 2.11a, each amide hydrogen in the protein is a tiny radio station broadcasting at its ringing frequency or chemical shift, which is approximately 500 MHz (Fig. 2.14). Similar to how AM630 plays the note middle-C by modulating its amplitude, the hydrogen nucleus "plays" the note of its attached nitrogen by modulating its amplitude at the nitrogen's frequency (i.e., ~50 MHz, Fig. 2.14). The result: a two-dimensional spectrum that shows which hydrogens are connected to which nitrogens (Fig. 2.11a).

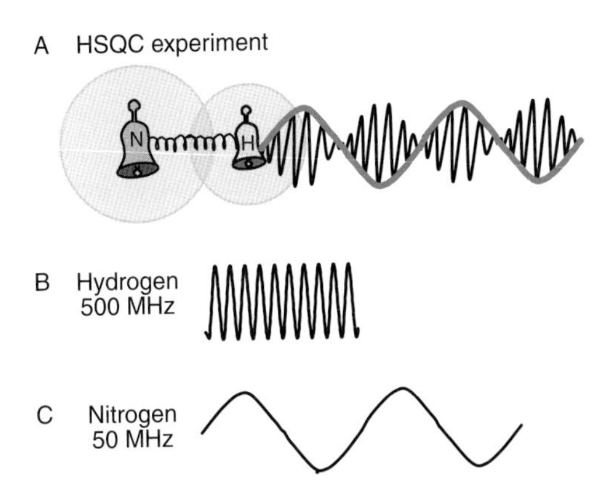

Fig. 2.14 (**a**) In a two-dimensional ^{1}H–^{15}N HSQC experiment, the hydrogen broadcasts its signal at (**b**) 500 MHz, but the wave's amplitude oscillates at the ringing frequency of the attached nitrogen, approximately (**c**) 50 MHz

How do you get an amide hydrogen to "play" the attached nitrogen's note? We use J-coupling! And, we run the experiment multiple times. Here's how it works.

In a two-dimensional NMR experiment, we start the nitrogen atom ringing; listen to its signal for a brief period; and then let it transfer magnetization to the bonded hydrogen via J-coupling (Fig. 2.10). We record the hydrogen's ringing as the NMR signal and then perform a Fourier Transform to determine the hydrogen's chemical shift.

Notice in step 2 of the HSQC experiment (Fig. 2.10) that the nitrogen's signal is a sine curve oscillating between a positive maximum value and negative minimum value. At some time points, the nitrogen's signal is even zero (arrows, Fig. 2.10).

To get the hydrogen's NMR signal to oscillate at the nitrogen's frequency, we merely run this experiment multiple times and vary the length of time the nitrogen is allowed to ring (Fig. 2.15). For example, during the first experiment, we let the nitrogen ring for the exact time that it takes for its electromagnetic wave to make one full cycle. At this point in time, the nitrogen's signal is close to its maximum value, so the amount of energy transferred to the hydrogen is maximal and the amplitude of the hydrogen's signal is large (experiment 1 in Fig. 2.15). In the second experiment, we let the nitrogen ring for a shorter amount of time, so that its signal is closer to zero. This time the amount of energy transferred to the hydrogen is significantly less than in the first experiment (experiment 2 in Fig. 2.15). This causes the hydrogen's signal to be smaller than in the first experiment. In the third experiment, we let the nitrogen ring so that its signal is almost zero when it transfers energy to the hydrogen. Then the amplitude of the hydrogen's signal is almost zero too (experiment 3 in Fig. 2.15). If we reduce the time period that the nitrogen rings even further (experiment 4 in Fig. 2.15), more energy is transferred to the hydrogen than in the previous experiment. Thus, the hydrogen's amplitude is larger in experiment 4 than in experiment 3.

The length of time that the nitrogen rings is called the t_1 *evolution period*. As we gradually shorten the t_1 period, you can see on the right-hand side of Fig. 2.15 that the amplitude of the hydrogen signal will

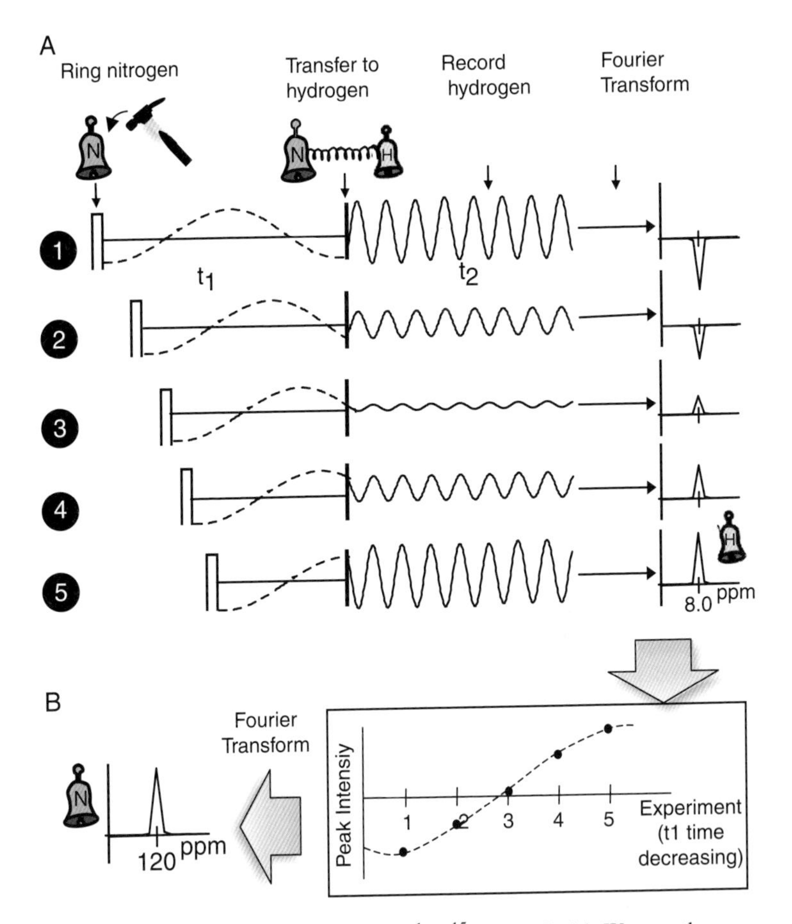

Fig. 2.15 Details of a two-dimensional ^1H–^{15}N HSQC. (**a**) We run the one-dimensional experiment multiple times, like the five experiments shown, varying the time (t_1) that the nitrogen atom is allowed to ring. This causes the intensity of the hydrogen peak to oscillate up and down at the nitrogen's frequency (graph inset). (**b**) If we apply a Fourier Transform to the plot of peak height versus t_1 time, we retrieve the ringing frequency of the nitrogen atom

oscillate up and down at the frequency of the nitrogen's signal—voila! The hydrogen's amplitude is changing at the frequency of its neighboring nitrogen (Fig. 2.15). In other words, the hydrogen is broadcasting the chemical shift of its neighboring nitrogen by amplitude modulation, similar to an AM radio station.

All that's left is to convert the two time signals into frequency signals, with the Fourier Transform. In this case, it's a double Fourier Transform. First, we perform a Fourier Transform on the hydrogen signal in each experiment. Notice that the resulting peak is at the same frequency in all experiments (the hydrogen frequency) but the height of the peaks varies (Fig. 2.15, far right). The curve created by these oscillating peaks encodes the frequency of the directly bonded nitrogen. To get this frequency, we perform the second Fourier Transform on the curve created by the oscillating hydrogen peak (Fig. 2.15b). That's it!

References and Further Reading

Cantor CR, Schimmel PR (1980) Biophysical chemistry part I: the conformation of biological molecules, techniques, Chaps. 2 and 5. W. H. Freeman and Company, New York.

Cavanagh J, Fairbrother WJ, Palmer AG III, Rance M, Skeleton NJ (2007) Protein NMR spectroscopy: principles and practice, 2nd edn., Chaps. 2, 4 and 7. Academic Press, Amerstdam.

Levitt M (2001) Spin dynamics: basics of nuclear magnetic resonance, Chaps. 7, 13, and 14. John Wiley & Sons, Inc., Chichester.

Wüthrich K (1986) NMR of proteins and nucleic acids, Chaps. 2–5. John Wiley & Sons, Inc., Chichester.

Chapter 3
Neighboring Bells and Structure Bundles

You know the old folklore: A portly soprano belts out a high note at the climax of an opera and BAM! Crystal chandeliers explode and champagne flutes shatter. Is this an operatic myth, or it is physically possible for the human voice to shatter glass?

In 2005, the American duo known as the "Myth Busters" (Jamie Hyneman and Adam Savage) put this glass-shattering fable to the test on their popular Discovery-channel television show. They hired Jamie Vendura, a voice coach and rock star, to attempt breaking a crystal wine glass. Although it took more than 20 attempts, Vendura could indeed shatter crystal by singing a note with a frequency matching the vibrational frequency of the glass (Fig. 3.1a). Further, when his voice was amplified through a speaker, the glass broke immediately every time.

What's happening here? Vendura's larynx creates sound waves that travel through the air to the crystal glass. When the frequency of the wave is near the natural frequency at which the glass vibrates, energy builds up and eventually cracks the glass (Fig. 3.1b).

Like Vendura's voice, the energy of ringing atoms can also travel through space and make neighboring atoms start to ring. This is called "through space energy transfer," and it tells us how close two atoms are to each other, even if the atoms are not directly connected by covalent bonds. We then use these distance measurements between many pairs

M. Doucleff et al., *Pocket Guide to Biomolecular NMR*,
DOI 10.1007/978-3-642-16251-0_3,

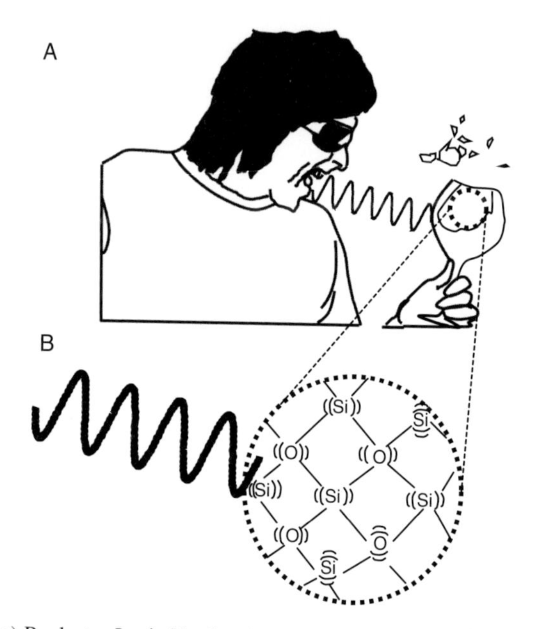

Fig. 3.1 (**a**) Rock star Jamie Vendura breaks a wine glass with his voice. (**b**) The energy travels through space from his larynx to the crystal, vibrating the atoms at their "natural" frequency

of atoms to build a three-dimensional structure of the molecule. Let's see how this works.

3.1 Bumping Bells: Dipole-Dipole Coupling

In the previous chapter, we learned that a ringing nucleus transfers its excess energy to nearby atoms that are connected by one or at most a few (up to three) covalent bonds. This useful phenomenon, called J-coupling, can be used to tell us which atoms are bonded to each other and helps us determine the chemical shift (or ringing frequency) for those atoms.

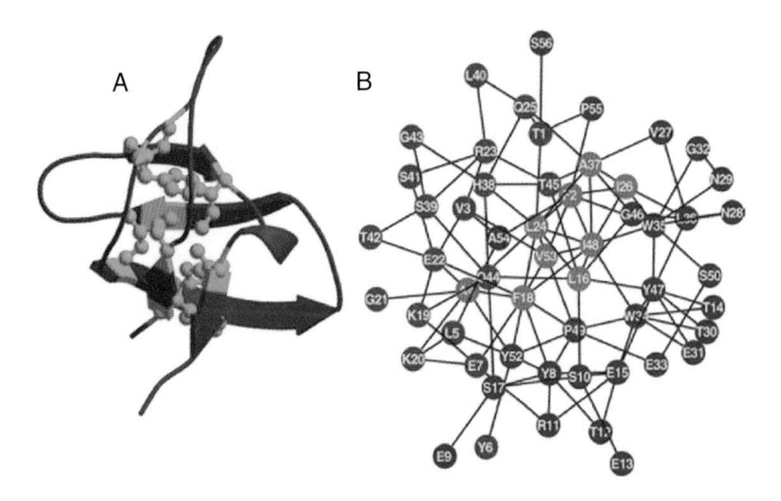

Fig. 3.2 (**a**) Ribbon diagram of the high-resolution structure of a small protein, the SH3 domain; residues participating in non-covalent, hydrophobic interactions are shown has small spheres. (**b**) Graph representation of the non-covalent interactions between residues in the SH3 domain. (PDB ID 1FMK; Lindorff-Larsen K et al. (2004) Nat Struct Mol Biol 11:443–449)

However, J-coupling is quite limited—it exists only between directly bonded or close indirectly bonded atoms and provides no information about atoms that are close in space but distant in the linear amino acid sequence of a protein. The three-dimensional structure of a protein depends heavily on these "non-covalent" interactions, especially between hydrophobic residues buried in the interior of a protein. For example, the small SH3 domain has only ~60 amino acids (Fig. 3.2a), but the hydrophobic residues in its interior form an extensive network of non-covalent interactions (Fig. 3.2b), which hold the structure together like glue. Therefore, to obtain a high-resolution image of a protein, such as calmodulin, we need a tool for measuring the distance between atoms that are *not* attached by bonds. That's where dipole-dipole coupling comes to our rescue!

Atoms don't need to be connected by a covalent bond to share excess energy or magnetization; they just need to be close in space. When two hydrogen atoms are specifically less than ~5–6 Å (5–6 × 10^{-10} m), they can transfer magnetization to each other through a process called "dipole-dipole" coupling (Fig. 3.3). Now 5 Å is a pretty

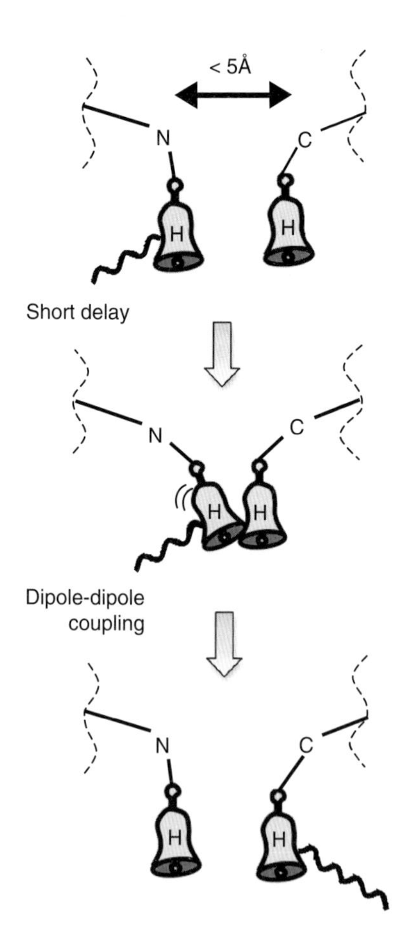

Fig. 3.3 A ringing atom transfers its excess energy via dipole-dipole coupling to neighboring atoms that are less than 5–6 Å apart

small distance—it's approximately 1/1000th the size of a bacterium or only 2.5 times the size of a chlorine atom. So, hydrogen atoms need to be quite close to share magnetization. But that's exactly what we want. If we know all the pairs of hydrogen in a protein that are less than 5 Å apart, we can definitely build a three-dimensional model for this complex molecule (Fig. 3.2a). We'll see how this "model-building" works in Sect. 3.6, but first, let's learn more about the magic of dipole-dipole coupling.

We'll start by comparing dipole-dipole coupling to Vendura's glass-shattering notes. In the "Myth Busters" episode, Vendura had to place his mouth right next to the glass surface to get the crystal wine glass to break. This is because the energy or intensity of our voice spreads out and weakens as you move away from the energy source. If you stand two feet from someone, their voice is four times softer than if you are only one feet from them (Fig. 3.4a). In other words, the intensity of their voice depends on $1/r^2$, where r is the distance between you and the sound source. Thus, the closer Vendura gets to the glass, the more energy transfers from his larynx to the molecules in the wine glass, and the faster he can break the glass.

Energy transfer via dipole-dipole coupling also depends greatly on the distance between the two atoms. The closer two atoms are in space, the more likely the atoms will exchange magnetization by dipole-dipole coupling. This "distance dependence" is even stronger than it is with sound—in dipole-dipole coupling, the chance of energy transfer depends on $1/r^6$, where r is the distance between the atoms (Fig. 3.4b and Mathematical Sidebar 3.1). If sound waves behaved this way, then standing two foot from someone would be 64 times softer than standing one feet from them (Fig. 3.4b)! Can you imagine? We would need a hearing aid at even the loudest rock concerts.

Thus, energy transfer by dipole-dipole coupling is much more sensitive to distance than sound transfer—the chance of a ringing atom sharing energy with its neighbor drops off very quickly as you move away from the high-energy atom (Fig. 3.4b). This is why dipole-dipole coupling occurs only between atoms that are less than ~5–6 Å apart; the effect is just too small for atoms farther away. But that's okay because

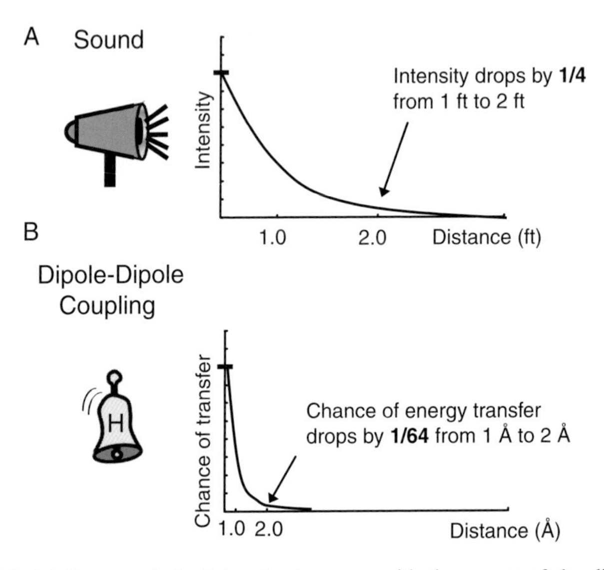

Fig. 3.4 (**a**) For sound, the intensity decreases with the square of the distance (intensity = 1/distance^2). (**b**) For dipole-dipole coupling, the probability of energy transfer decreases with the distance raised to the sixth power (probability = 1/distance^6)

for many biological molecules, such as proteins, all we need to know is which atoms are close to each other.

Although dipole-dipole coupling and sound have similar distance dependencies, the two energy transfer mechanisms are significantly different. For starters, if you rub your finger around the rim of crystal wine glass, the glass hums at almost a single frequency. This is called the "natural frequency" of the crystal, and to shatter the wine glass, Vendura had to sing a note close to this frequency.

Does this same idea hold true for dipole-dipole coupling between atoms? Surprisingly, no. The frequency of the ringing atom does not need to match the "natural" frequency of the neighboring atom to share energy via dipole-dipole coupling.

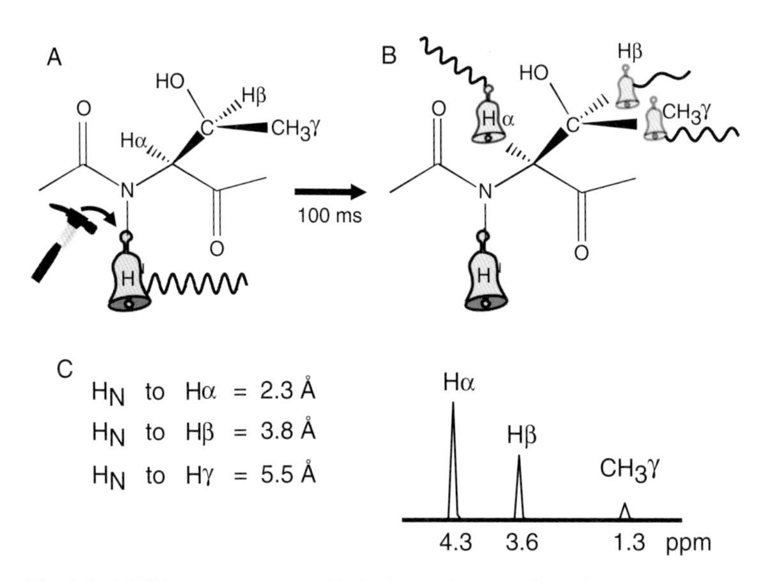

Fig. 3.5 (**a**) When we ring an amide hydrogen in a protein and then wait a short period of time, (**b**) other hydrogens in the protein, such as the Hαs, Hβs, and Hγs, may start ringing and produce peaks in the NMR spectrum; (**c**) the height of these peaks depends on the distance between the two hydrogen atoms

For example, in the valine–threonine peptide, the amide hydrogen (H_N) (i.e., the hydrogen atom attached to the nitrogen in Fig. 3.5a) rings at 8.5 ppm, and the nearby α-hydrogen (H_α) rings at 4.3 ppm, a difference of more than 2,000 Hz on a 500 MHz NMR spectrometer. However, if we ring the H_N (Fig. 3.3) and wait a short time, like 50–100 ms, we can hear the Hα start ringing too. This is because the two atoms are only ~2.3 Å apart and are very likely to exchange energy by dipole-dipole coupling. The H_N can even give its excess energy via dipole-dipole coupling to the neighboring nitrogen and carbon atoms even though their ringing frequencies (~50–125 MHz respectively) are more than four- to tenfold smaller than that of hydrogen's (~500 MHz)! (If you've had experience with quantum mechanics, this seems counterintuitive. But we'll see in the next chapter that the matching frequency comes from the motion of the molecule in combination with the nucleus's ringing frequency.)

The second major difference between energy exchange by dipole-dipole coupling and breaking glass with your voice is the time it takes to transfer the energy. When Vendura begins singing, it takes time for the sound waves to travel through the air and hit the atoms in the wine glass. This time is very small, less than a millisecond, but there is a moment where you can detect the sounds waves before they start rattling the atoms. This type of energy transfer is called "radiative" because the energy source "radiates" a detectable wave for the target to adsorb (Fig. 3.1a).

In contrast, energy transfer by dipole-dipole coupling is non-radiative. The high-energy nucleus doesn't emit a wave for the neighboring atom to adsorb; instead, when the nucleus decides it's the right time, it spontaneously gives its extra energy to a nearby nucleus, so it can start ringing. The transfer is instantaneous, and there is no detectable electromagnetic wave between the two atoms.

In this way, non-radiative energy transfer by dipole-dipole coupling is more like two bells "bumping" each other, exchanging energy instantaneously without the need for an intermediate wave between them (Fig. 3.3). Or, think of two billiard balls colliding on a pool table. When the moving ball hits a still ball, almost all the kinetic energy is immediately transferred to the second ball—the same is true for two atoms in dipole-dipole coupling.

How does the ringing nucleus "decide" to bump into a neighboring nucleus? There are beautiful mathematical equations to describe this decision, and we'll learn more about it in the next chapter. But for now, the most important factor is the distance dependency—the closer the two atoms are, the greater the chance for the bells or atoms to "collide" (Fig. 3.4b). This distance dependence makes dipole-dipole coupling one of the most powerful techniques scientists have for measuring how far apart two atoms are in space. Let's see how bio-NMR spectroscopists use this atomic measuring stick.

Mathematical Sidebar 3.1: Dipole-Dipole Coupling

Dipolar relaxation occurs between two spins: one that is creating the magnetic field (I) and one that is experiencing it (S). The strength of this field (d) depends on the distance between the two spins (r), each spins gyromagnetic ratio (γ_I and γ_S) and the orientation of the vector between the two spins relative to the magnetic field (B_o). Putting these together, we get a proportionality that looks like

$$d \propto \gamma_I^2 \gamma_S^2 r_{IS}^{-6}$$

Clearly, the energy transfer by dipole-dipole coupling decreases rapidly as you separate the two coupled nuclei. This proportionality also explains why dipole-dipole coupling is much more efficient for protons than for ^{13}C or ^{15}N atoms, which have much smaller gyromagnetic ratios than protons.

3.2 Atomic Meter Stick: The NOE

Energy transfer via dipole-dipole coupling is called the "Nuclear Offerhauser Effect," or NOE, named after the Californian physicist Albert Overhauser, who first predicted the phenomenon while he was a postdoctoral fellow at the University of Illinois. When Dr. Overhauser presented his unique idea of energy transfer to the American Physics Society in 1953, the top physicists in the crowd, such as Felix Bloch (1952 Physics Nobel Prize), Edward M. Purcell (1952 Physics Noble Prize), and Norman F. Ramsey (1989 Physics Noble Prize) were taken aback and quite skeptical. Dr. Ramsey even wrote Dr. Overhauser a letter a few months after the meeting (Overhauser, 1996):

Dear Dr. Overhauser:

> You may recall that at the Washington Meeting of the Physical Society, when you presented your paper on nuclear alignment, Bloch, Rabi, Purcell, and myself all said that we found it difficult to believe your conclusions and suspected that some fundamental fallacy would turn up in your argument. Subsequent to my coming to Brookhaven from Harvard for the summer, I have had occasion to see the manuscript of your paper.
>
> After considerable effort in trying to find the fallacy in your argument, I finally concluded that there was no fundamental fallacy to be found. Indeed, my feeling is that this provides a most intriguing and interesting technique...

Turns out that energy transfer by dipole-dipole coupling was not obvious to physicists because it is a "second order effect"—in other words, it required an extra bit of math to predict it. The "primary effect" of dipole-dipole coupling doesn't show up in solution because molecules are tumbling around in all directions, randomizing the effect. So physicists weren't too interested in studying it in solution. But when Dr. Overhauser predicted this extraordinary energy transfer mechanism, physicists went hunting for it. In less than a year, Ionel Solomon, a physicist at Harvard University, confirmed Dr. Overhauser's predictions by showing that the NOE does indeed exist in solution. Sixty years later, the NOE, or energy transfer by dipole-dipole coupling, is the most important technique for determining the distance between protons in solution. Still today, the NOE provides the mainstay for three-dimensional structure determination of proteins in solution.

You can think about the NOE as a measuring stick (Fig. 3.6) that tells us how far apart two protons are in a structure. But this "NOE stick" is not your typical measuring device. First, it only extends to 5–6 Å (Fig. 3.6); if the two protons are farther apart, the NOE is generally so small that one cannot measure their separation by dipole-dipole coupling. Second, in practice this NOE stick is best interpreted in terms of broad ranges: 1.8–2.5 Å, 1.8–3.5 Å, and 1.8–6 Å, where the lower boundary is simply the sum of the van der Waals radii of two protons (i.e., they cannot get any closer together) and the upper boundary is the limit of detection of the dipole-dipole interaction (Fig. 3.6). Also,

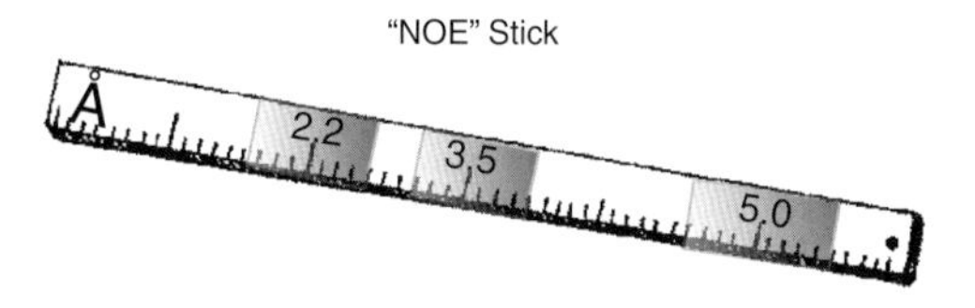

Fig. 3.6 Dipole-dipole coupling (or NOE) is similar to a measuring stick with poor precision and only a few demarcations (2.2, 3.5, 5.5 Å)

notice the notches on the "NOE stick" are quite fuzzy because with dipole-dipole coupling we can say only that two atoms are *at most* ~2.5 Å, or at most ~3.5 Å apart, or at most ~5–6 Å apart. In other words, distance measurements by the NOE are typically only semi-quantitative; it usually gives us only a rough estimate of the *maximum* distance between two protons. This is because other phenomena, which we'll learn about in the next chapters, can interfere with the energy transfer via dipole-dipole coupling.

Let's look at the threonine residue in the peptide again (Fig. 3.5a). In the three-dimensional structure, the threonine's H_N atom is 2.3 Å from the Hα, 3.8 Å from the Hβ hydrogens, and 5.5 Å from the Hγ hydrogen. If we start ringing the H_N hydrogen (Fig. 3.5a), wait for a bit of time to allow dipole-dipole coupling to occur (say 100 ms), and collect the NMR signal from all the hydrogens. What would we see?

Before we answer that, we need to learn one more idea about NMR spectroscopy of big, chunky molecules, such as proteins and oligonucleotides. Atoms ring incredibly softly, or more accurately the electromagnetic wave they create is extremely weak. The only way we can detect their ringing is to have a huge quantity of atoms all ringing together. Most NMR samples contain about 10^{16} molecules (hundreds of micromolar in 0.5 ml). When we ring the H_N hydrogen in the valine–threonine peptide, all 10^{16} H_N atoms start ringing together, like a bell choir with millions of tiny bells. It's actually quite amazing when you think about it! (We'll talk more about this fantastic feat when we learn about coherence, but let's get back to dipole-dipole coupling.)

So we start ringing all 10^{16} H_N protons (Fig. 3.5a) and then wait a brief period of time. What happens during this delay? Most H_N atoms in the sample will keep all their excess energy and continue to ring. But some H_N atoms will hand off their extra magnetization to a nearby hydrogen by dipole-dipole coupling. This extra energy allows the neighboring atom to start ringing and produce an NMR signal (Fig. 3.5b).

These nascent signals from nearby atoms create cross-peaks in the NMR spectrum. And, the relative peak height tells us approximately how close they are to the H_N atom that gave them the energy. (Remember that the cross-peak height depends directly on the strength of the NMR signal or how many atoms are ringing.) The $H\alpha$ is the closest to the H_N, so its cross-peak is by far the strongest (Fig. 3.5c). In contrast, the $H\gamma$s are 5.5 Å from the H_N hydrogen, so they rarely get sent energy from the H_N, resulting in a barely detectable cross-peak (Fig. 3.4c). Notice how the $H\alpha$ is half the distance from H_N (2.8 Å) as the $H\gamma$ is from H_N (5.5 Å). However, the $H\alpha$'s cross-peak is much stronger (~64 times the intensity!) than the $H\gamma$'s cross-peak in the 2D NMR spectrum (Fig. 3.5c). This is just as we would predict given that energy transfer by dipole-dipole coupling or the NOE depends on the inverse sixth power of the distance ($1/r^6$).

In Chap. 2, we used the J-coupling energy transfer mechanism to create a two-dimensional NMR spectrum with the *x*-axis containing chemical shift (or ringing frequency) of the amide hydrogens, and the *y*-axis giving the chemical shift of the nitrogen covalently bound to the nitrogen. We can use the NOE to create a similar type of two-dimensional spectrum (Fig. 3.7a):

1. Ring all the amide hydrogens.
2. Record the chemical shifts.
3. Allow hydrogens to transfer energy to neighboring hydrogens via the NOE.
4. Record the NMR signal of all ringing hydrogens.

The result is a spectrum called a 2D NOE (Fig. 3.7b), and it shows which hydrogens are *close to each other in space*. We read the 2D

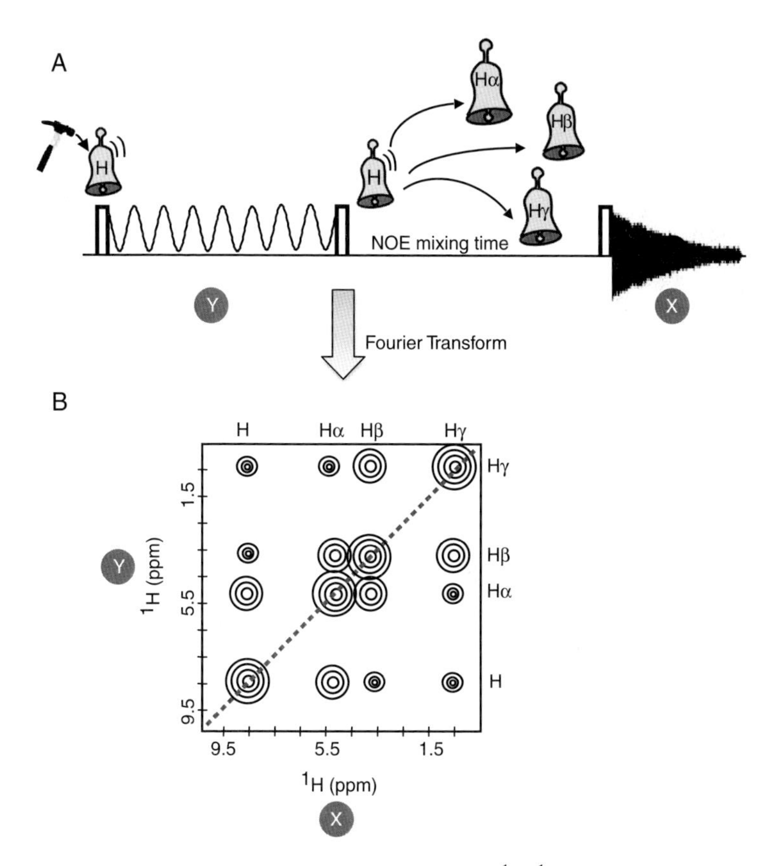

Fig. 3.7 (**a**) Pulse sequence for the two-dimensional ^1H–^1H NOESY experiment, (**b**) which creates a two-dimensional spectrum with a diagonal peak (dotted line) for each hydrogen atom and a cross-peak for each pair of hydrogen atoms that are less than 5–6 Å apart

NOE just like we read the 2D HSQC except now both axes give the chemical shifts for hydrogens in the molecule that are <5–6 Å from each other. For example, in the cartoon of a 2D NOE (Fig. 3.7b), the amide hydrogen (H_N) for the threonine residue has a frequency of

9 ppm. If we draw a vertical line at $x = 9$ ppm, we see four cross-peaks: one at 9 ppm, the cross-peak to itself (see below for explanation of this); one at 5.2 ppm, the cross-peak to the Hα; one at 3.5 ppm, the cross-peak to the Hβ; and one at 1.0, the cross-peak to the Hγ. Notice how the size of the cross-peaks tells us the distance between the two hydrogens. For example, the cross-peak between the H_N and the Hγ is substantially smaller than the one between the H_N and H_α because H_N and H_α are much closer to each other than the H_N and the Hγ atoms.

Look at the peaks along the dotted line, for example, at $y = 9.0$ ppm and $x = 9.0$ ppm. Notice that this peak is quite large. Where does this super peak come from?

In the last section we learned that ringing atoms transfer only part of their excess energy; actually most of their energy they keep for themselves during the NOE mixing time. This "kept" energy creates strong peaks at $x = y$ for all amide protons in the protein, producing a streak across the 2D-NOE spectrum (Fig. 3.7b, dotted line). Not surprisingly, these peaks are called the diagonal (or "auto-peaks"). Diagonal peaks have wonderfully large signals, but unfortunately they don't contain any useful geometric information because we already know that the H_N is close to itself!

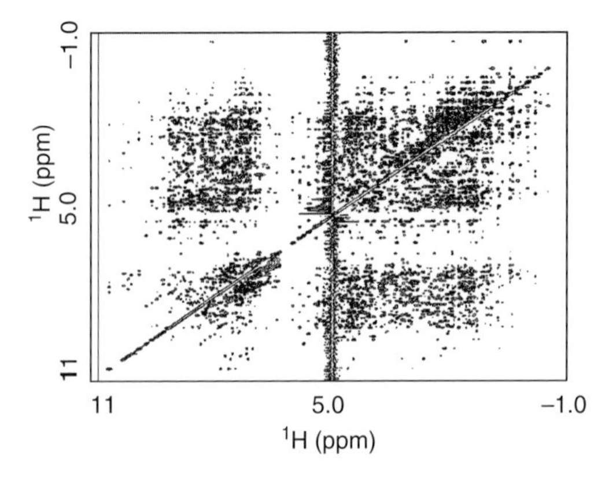

Fig. 3.8 Two-dimensional ^{1}H–^{1}H NOESY spectrum recorded on a small protein (15 kDa)

In contrast, the cross-peaks away from the diagonal are rich with structural information. Unfortunately, for most proteins, including average size proteins like calmodulin, the 2D-NOE spectrum is usually too crowded and overlapped to be useful for structure determination. For example, check out the 2D-NOE spectrum for lysosyme (Fig. 3.8), which has only ~130 amino acids! How can we fix that? You got it—add another dimension.

3.3 Into "Three-D"

In Chap. 2, we saw how adding an extra dimension can greatly simplify our NMR spectrum by spreading out overcrowded peaks and selecting for only specific atoms. So far we've learned about two types of 2D NMR experiments:

(a) The HSQC, which uses J-coupling to tell us what atoms are connected by covalent bonds.
(b) The NOE, which uses dipole-dipole coupling to show us which atoms in a molecule are close in space.

Two-dimensional experiments similar to these two types are all you basically need to solve the structure of small protein or DNA molecule that is less than 5–10 kDa. But these experiments need to be stepped up a notch to be useful for larger biomolecules. Turns out we can combine the HSQC and NOE experiments together to create a 3D-HSQC-NOE experiment. It's rather simple—we merely run the NOE part and then immediately run the HSQC experiment (Fig. 3.9a). Here's how it would go[1]:

---

[1] Again, this is a simplification of the actual experiment, which you can find in Chap. 7 of Cavanagh et al. (2007). Most notably, three-dimensional experiments typically start by ringing hydrogen atoms because hydrogens have the largest gyromagnetic ratio. This then requires an extra step in the experiment to record the nitrogen chemical shifts and return back to amide hydrogens before the NOE transfer.

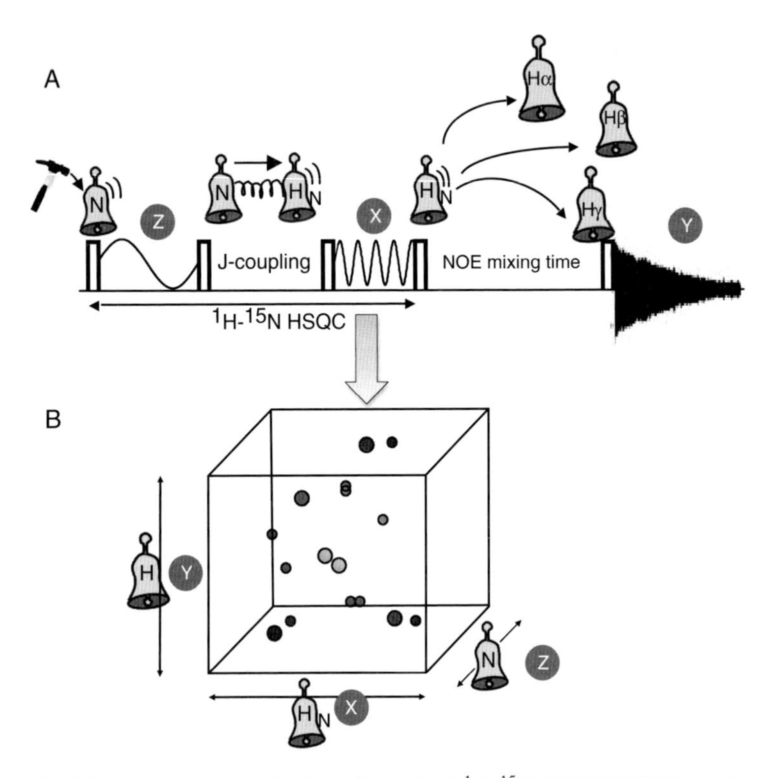

Fig. 3.9 (**a**) Sequence for the three-dimensional ^{1}H–^{15}N-HSQC-NOESY experiment, which records the ringing frequency of a nitrogen atom, its covalently linked hydrogen, and every hydrogen <5 Å from the amide hydrogen. (**b**) This creates a cube of data with one peak for each trio of atoms (H_N, N, and H_X)

1. Start ringing the nitrogen atoms and record their chemical shift.
2. Use J-coupling to transfer magnetization to amide hydrogens via J-coupling (the HSQC part).
3. Let the amide hydrogens ring a bit and record their chemical shifts.
4. Let the amide hydrogens transfer magnetization to nearby hydrogens via dipole-dipole coupling (the NOE part).
5. Record the NMR signals of the protons.

Instead of the plane of data that we saw in the 2D experiments, this 3D experiment produces a *cube* of data (Fig. 3.9b). Each side of the cube (or axis) gives the chemical shift for one of the three atoms recorded in the experiment: the first dimension (*y*-axis) gives the ringing frequencies for all protons, the second dimension (*z*-axis) provides the frequencies for the amide nitrogens, and the third dimension (*x*-axis) gives the ringing frequency of the amide protons ≤5 Å from the protons labeled on the *y*-axis.

The easiest way to analyze this spectrum is to divide it up into two parts:

1. The 2D-HSQC plane (Fig. 3.10a), which projects all the peaks into a *z*–*x* plane and looks identical to the 2D-HSQC that we learned about in Chap. 2 (compare Fig. 3.10a with Fig. 2.8b).
2. NOE strips, which contain all peaks along the *y*-axis for a given peak in the HSQC spectrum (Fig. 3.10b).

Let's see how it works. Say we want to find all the hydrogens <5 Å from the amide hydrogen of residue 137 in calmodulin (i.e., 137-H_N).

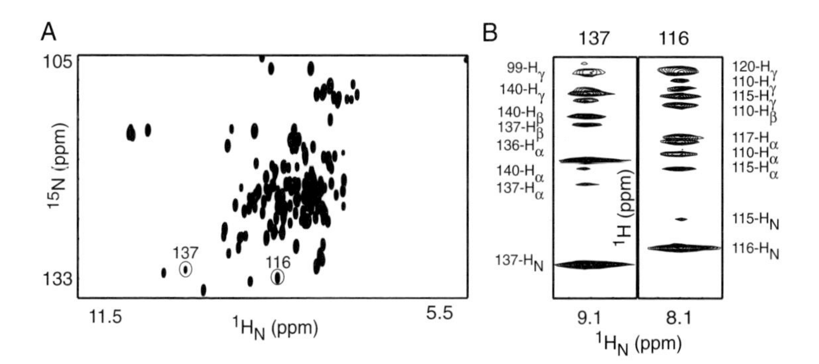

Fig. 3.10 (**a**) HSQC plane of the 3D-HSQC-NOESY recorded on the protein calmodulin (17 kDa). (**b**) ^1H–^1H 2D-NOE strips ("the *y*-axis") for the residues H_N-137 and H_N-116 (circled in (**a**)); chemical shift assignments are shown on the sides

Furthermore, we know that it has a chemical shift of 9.104 ppm, and the nitrogen attached to it (i.e., 137-N) has a chemical shift of 131.1 ppm. We simply go to $x = 9.104$ and $z = 131.1$ in the 2D-HSQC plane (Fig. 3.10a) and then "look" down the y-axis or the 2D-NOE "strip". Each cross-peak in the 2D strip represents a hydrogen near 137-H_N. For example, 137-H_N is very close to the Hγ of residue 140, so the cross-peak for this pair of hydrogens is quite strong. Check out the cross-peak at the top of the strip that is labeled 99-Hγ. Although residue 99 is very far from residue 137 in the sequence of calmodulin, in the three-dimensional structure these two residues are quite close to each other (<5 Å). Thus, 137-H_N and 99-Hγ create a clear, although weak cross-peak in the 3D-HSQC-NOE. As we'll see later in this chapter, these "long range" NOE cross-peaks between atoms that are distant in the protein sequence but close together in the three-dimensional structure are paramount for building a high-resolution three-dimensional structure.

For atoms that are well separated in the 2D-HSQC spectrum, such as residue 137 in calmodulin (Fig. 3.8), adding the third dimension doesn't help much. However, many of the amide hydrogens are super crowded in the 2D-HSQC, and an extra dimension is essential for obtaining distance information for these atoms.

Let's look at an example. Say the amide hydrogen for residue 116 (116-H_N) has a chemical shift of 8.114 ppm. If we draw a vertical line passing through $x = 8.114$ ppm in the 2D-HSQC plane (Fig. 3.10a), we find at least *seven* other amide hydrogen cross-peaks along this line, in addition to the one for residue 116. This means that in the regular 2D-NOE spectrum, all the cross-peaks at 8.114 ppm arise from hydrogens physically near residue 116 *and* physically near the other *seven* amide hydrogens with a chemical shift of 8.114 ppm. How can we differentiate which cross-peaks belong to 116-H_N and which ones belong to other hydrogens? This is where the 3D-HSQC-NOE really helps us out.

In the 3D-HSQC-NOE spectrum, we merely go to H_N-116's cross-peaks in the HSQC plane (the x–z plane or ^{1}H–^{15}N plane) and look down the y-axis. The NOE "strip" from the 3D experiment gives us the cross-peaks for only the hydrogens near H_N-116. All the cross-peaks for the other amide hydrogens at 8.114 ppm are conveniently absent! (Actually, they are in strips located at different points along the *y*-axis in the HSQC plane or on different *x*–*z* planes). Now we can easily tell which hydrogens are close to H_N-116 and determine their cross-peaks' heights to estimate their distance from H_N-116. Thus, the 3D spectrum is much easier to read than the messy 2D-NOE spectrum (like the one shown in Fig. 3.8). Thank goodness for 3D spectra!

Plus, this 3D-HSQC-NOESY is just the tip of the 3D spectra iceberg. There are dozens of three-dimensional experiments we can create by combining simple HSQC and NOE experiments. And if we need even more resolution we can simply extend the number of dimensions to four.

The 3D experiment we've been discussing (Figs. 3.9 and 3.10) is specifically called the 3D-^{15}N-separated NOE because we select for protons attached to nitrogen atoms in the HSQC part of the experiment. What about the hydrogens attached to carbon atoms?

To get distance information for these protons, all we need to do is ring the carbon atoms instead of the nitrogen atoms at the beginning of the experiment (Fig. 3.9a).[2] The result is a spectrum called 3D-^{13}C-separated NOE, and it is very similar to the ^{15}N-separated NOE spectrum except now the *x*-axis provides the chemical shift for all hydrogens attached to carbon atoms.

Turns out that this 3D-^{13}C-separated NOE spectrum provides the key distance information for structure determination by NMR because most proteins are "held together" by interactions between hydrophobic hydrogens covalently bonded to carbons. We'll see how this works soon, but before we move on, we need to answer one more question: how did we know that the $H\gamma$-116 hydrogen has a ringing frequency of 8.114 ppm? And, how did we know how to label all those cross-peaks in Fig. 3.10b?

---

[2] In practice, we select the carbon atoms by setting the J-coupling delay in the HSQC to optimize energy transfer between carbons and hydrogens instead of transfer from nitrogens and hydrogens. Again, see Chap. 7 of Cavanagh et al. (2007) for details.

3.4 Adult "Connect-the-Dots:" HNCA

As you probably guessed from the previous section, a NOE spectrum, whether it is 2D, 3D, or 4D, is useful for structure determination only if we know the ringing frequencies for nearly all the hydrogens, carbons, and nitrogen atoms in the protein. For a decent size protein (>10 kDa), this is literally thousands of atoms! For example, the 148-residue calmodulin protein has 1,129 hydrogens, 719 carbons, and 189 nitrogens. How on Earth do we figure out all these chemical shifts?

The process of determining the ringing frequencies for all atoms in a molecule is called "chemical shift assignment." It can be a tricky matter, especially for larger proteins (>20 kDa). But for all molecules, the idea is the same. We string together different types of HSQC experiments and record the NMR signal for specific atoms as we hop through the bonds of the molecule via J-coupling. The result is a 3D spectrum telling you which chemical shifts are connected by covalent bonds. Once you know the chemical shifts of a few atoms in a protein, you can start "linking" the cross-peaks together just like the amino acids are linked together in the sequence. It's probably easiest to look at an example to see how it works.

One of the most common 3D experiments used to assign the chemical shifts of a protein's backbone atoms (the H_N, N, and $C\alpha$ atoms) is called the 3D HNCA, and it's a great place to start solving a structure. Here's how it goes (Fig. 3.11):

1. Ring the H_N atoms and then transfer the magnetization to the attached nitrogen via J-coupling to the bonded nitrogens (this is just a regular $1H–^{15}N$ HSQC).
2. Record the nitrogen's chemical shift.
3. Then transfer its excess magnetization to the neighboring $C\alpha$ atoms via J-coupling.
4. Record the $C\alpha$ chemical shift.
5. Then go BACK to the amide hydrogen and record its NMR signal.

The result is a cube of data similar to the 3D-^{15}N-separated NOE above but with the following axes (Fig. 3.13b):

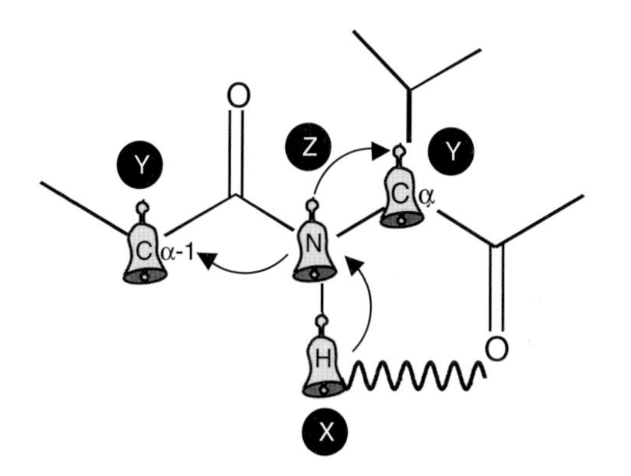

Fig. 3.11 In the 3D-HNCA experiment, excess energy from the ringing amide hydrogen (H_N) is transferred to the attached nitrogen atom (N) via J-coupling and then to both the Cα and the Cα-1 atoms. This creates a cube of data, like Fig. 3.9b but with the Cα chemical shifts on the *y*-axis instead of N chemical shifts

x = chemical shifts of amide hydrogens
z = chemical shifts of amide nitrogens
y = chemical shifts of Cα atoms

Notice that the *x*–*z* plane is merely a regular ^{1}H–^{15}N HSQC with H_N and N chemical shifts on each axis, exactly as it was in the 3D-^{15}N-separated NOE (compare Fig. 3.12A with Fig. 3.10a). It's the *y*-axis where the magic happens! For example, say we already know where residue 137's cross-peak is in the ^{1}H–^{15}N HSQC, and we want to find the cross-peak for residue 136. We merely go to residue 137's peak in the *x*–*z* plane (Fig. 3.12a: x = 9.104 ppm; z = 131.1 ppm) and then "look" down the *y*-axis, where we'll find cross-peaks for the residue's Cα atom (Fig. 3.13b). But why does this 2D strip have two cross-peaks—a strong one and a weaker one at 71 and 55 ppm, respectively? There is only one Cα covalently attached to the amide nitrogen in Fig. 3.11, so why are there two peaks in this spectrum?

Interestingly, the J-coupling constants are almost the same for the one-bond J-coupling between 137-N and 137-Cα (~9 Hz) and the

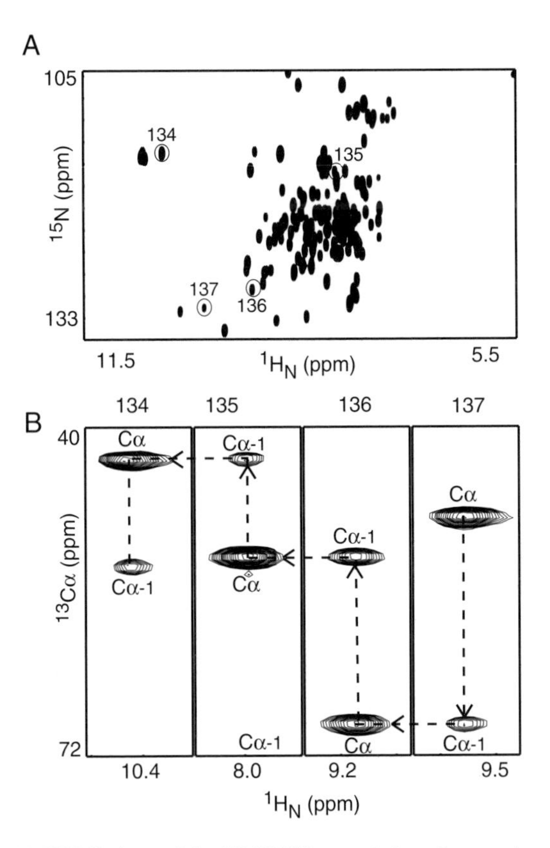

Fig. 3.12 (**a**) HSQC plane of the 3D HNCA recorded on the protein calmodulin (17 kDa). (**b**) ^{1}H–^{1}Cα 2D strips ("the *y*-axis") for the residues circled in (**a**); each residue produces a cross-peak for both the residue's Cα and the previous residue's Cα (Cα-1)

two-bond J-coupling between 137-N and 136-Cα, (i.e., the Cα on the other side of the peptide bond in residue 136) (~7 Hz). This means that in the 3D-HNCA experiment, magnetization will go to both 137-Cα and 136-Cα when we transfer energy by J-coupling from the nitrogen

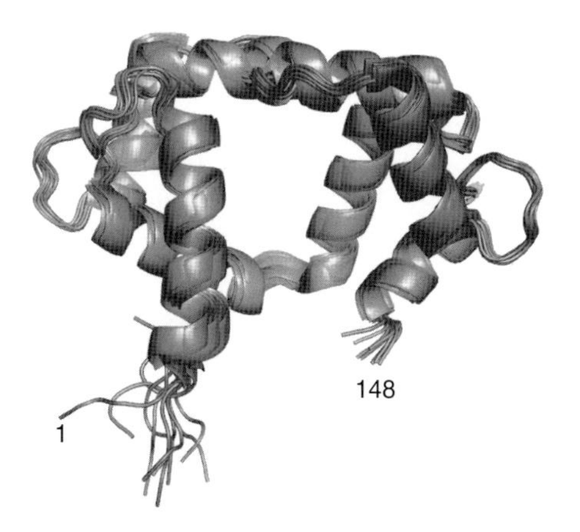

Fig. 3.13 NMR solution structure of calmodulin in a compact or closed configuration with the first and last amino acids labeled (ten lowest energy structures overlayed; PDB ID 1SY9)

(Fig. 3.11). Because 137-N has a slightly smaller J-coupling constant with 136-Cα than with 137-Cα, the cross-peak for 136-Cα is slightly weaker than the cross-peak for 137-Cα (Fig. 3.12b).

In general, this is true for each residue in the protein: The one-bond J-coupling between the nitrogen and its own Cα (~7–11 Hz) is very similar to the two-bond J-coupling between the nitrogen and the Cα in the previous residue (~4–9 Hz) (Fig. 3.11). So in the 3D HNCA, we see two cross-peaks along the *y*-axis for each amide hydrogen, one strong peak arising from the Cα covalently attached and one weak peak arising from the Cα in the previous residue (Fig. 3.12b).

The cool part about having "double" Cα peaks in the 3D HNCA is that we can use the Cα-136 peak to find the H_N and N chemical shifts for residue 136. We merely search for a 2D strip in the HNCA that has a strong cross-peak at 55 ppm (the same chemical shift as the weak cross-peak in 137's 2D strip) (Fig. 3.13b). The H_N and N chemical shifts for this strip tells us exactly where residue 136 is found in the ^{1}H–^{15}N HSQC (Fig. 3.12a).

We can repeat this process for residue 136, then 135, then 134, etc. Eventually we can assign the chemical shifts for all the H_N, N, and Cα atoms in the protein. As you can see from Fig. 3.12b, this linking up of the cross-peaks in the 3D HNCA turns into a gigantic "connect-the-dots." It's a bit tougher than it looks here because peaks often overlap or can be missing, but it's actually quite fun and challenging!

Once you have completed connecting the cross-peaks and assigning the ringing frequencies for the H_N, N, and Cα atoms in the 3D-HNCA spectrum, you can use other 3D J-coupling experiments to assign the atoms in the protein's sidechains. For example, the CBCA(CO)NH experiment transfers magnetization from the Cβs to the Cαs and then through the carbonyls to the amide hydrogens. So it gives you the chemical shifts for all the Cβ atoms. Analogously, the HBHA(CO)NH transfers magnetization from the Hβs and the Hαs through the carbonyls to the amide hydrogens. So it gives you the chemical shifts for Hβ atoms. You probably get the idea—the experiment's name tells you how the magnetization flows and what cross-peaks will be present in the spectra.

Can you draw a "pulse sequence diagram" for the CBCA(CO)NH, like the one given in Fig. 3.7 for 2D-NOE experiment? Can you guess what cross-peaks are present in CC(CO)NH experiment? What coupling constants would be used in this experiment?

3.5 Putting the Pieces Together: A Quick Review

We have made it! We now have all the ideas, experiments, and tools we need to start building a three-dimensional model of a protein. Before we start hammering and nailing the protein into shape, let's review the steps using the 17 kDa calmodulin protein as an example:

(a) First, you need to prepare a sample of highly pure (≥95%) calmodulin protein and concentrate it to about 0.5–1 mM (remember we need a high concentration of protein because each atom rings *very* softly). Also, the calmodulin needs to have all its ^{14}N atoms replaced with ^{15}N atoms and its ^{12}C atoms replaced with ^{13}C atoms (see Mathematical Sidebar 2.2 for why this step is needed).

(b) Now we run a regular 2D ^{1}H–^{15}N HSQC spectrum to check that the protein is "behaving well"—it's well folded, not aggregating, and generally going to be easy to work with. Remember you want all the cross-peaks in the ^{1}H–^{15}N HSQC to have comparable intensities and be reasonably well dispersed. In addition, the number of peaks should be approximately the same as the number of residues in the protein. Although these traits are usually needed to determine a protein's structure by NMR, "ill-behaved" proteins can also be studied by NMR to determine if they are folded, if they are aggregating or oligomerizing, and many other interesting biophysical properties.

(c) Now we run 3D-heteronuclear J-coupling experiments to start determining the ringing frequency for each hydrogen, carbon, and nitrogen in calmodulin. The exact experiments you need to perform varies from protein to protein, but you will definitely need ones like the 3D HNCA and the 3D CBCA(CO)NH.

(d) Once you've assigned all the cross-peaks in these experiments to atoms in the protein (i.e., "connected-the-dots"), you need to create a chemical shift table listing all the atoms in the protein (except oxygens and sulfurs) and their corresponding chemical shift value, as shown in Table 3.1.

(e) Now we use the chemical shift table to assign the cross-peaks in the 3D ^{13}C- and ^{15}N-separated NOE experiments, like the strips shown in Fig. 3.10c.

(f) Lastly we convert the NOE assignments into a list of approximate inter-proton distance restraints for the protein. Remember, each cross-peak in the NOE spectra is created by a pair of hydrogens that are <5 Å apart and the cross-peak intensity estimates the distance between the two atoms. For example, the NOE strip of residue 116 shown in Fig. 3.10b has eight cross-peaks, so it will give us eight inter-proton distance restraints. Generally, these are classified into three ranges, 1.8–2.7, 1.8–3.5, and 1.8–5 (or 6) Å, corresponding

Table 3.1 Example of a chemical shift assignment table

1428 136	ASP NH H 8.05 0.05
1429 136	ASP HA H 4.24 0.05
1430 136	ASP HB2 H 2.77 0.05
1431 136	ASP HB3 H 2.62 0.05
1432 136	ASP CO C 178.68 0.10
1433 136	ASP CA C 55.04 0.10
1434 136	ASP CB C 39.78 0.10
1435 136	ASP NH N 120.73 0.10
1436 137	THR NH H 9.10 0.05
1437 137	THR HA H 4.49 0.05
1438 137	THR HB H 4.80 0.05
1439 137	THR HG2 H 1.35 0.05
1440 137	THR CO C 175.61 0.1
1441 137	THR CA C 71.00 0.1
1442 137	THR CB C 72.05 0.1
1443 137	THR CG2 C 21.90 0.1
1444 137	THR N N 131.38 0.1

The columns from left to right are atom number, residue number, residue type, atom type, nucleus type, chemical shift (ppm), and error in chemical shift (ppm).

Table 3.2 Example of a NOE table

Assign (resid 137 and name HN) (resid 99 and name Hγ) 5.00 1.80
Assign (resid 137 and name HN) (resid 140 and name Hγ) 2.70 1.80
Assign (resid 137 and name HN) (resid 140 and name Hβ) 3.50 1.80
Assign (resid 137 and name HN) (resid 136 and name Hα) 2.70 1.80
Assign (resid 137 and name HN) (resid 140 and name Hα) 5.00 1.80
Assign (resid 116 and name HN) (resid 115 and name HN) 5.00 1.80
Assign (resid 116 and name HN) (resid 115 and name Hγ) 2.70 1.80
Assign (resid 116 and name HN) (resid 115 and name Hα) 3.50 1.80
Assign (resid 116 and name HN) (resid 110 and name Hβ) 2.70 1.80

Each line provides the name of two atoms and the maximum and mininum distance between them. For example, row one says "the amide hydrogen of residue 137 and the Hγ of residue 99 are at most 5.00 Å apart."

to strong, medium, and weak cross-peak intensities, respectively. For an average size protein like calmodulin, the 3D ^{13}C- and ^{15}N-separated NOE experiments should supply 1,000–2,000 distance restraints, such as the ones shown in Table 3.2.

3.6 Wet Noodles and Proteins Bundles: Building a Three-Dimensional Structure

Now we can start building the protein! At this point the hard work is basically behind us. All we do now is to give the list of “NOE restraints” (Table 3.2) to a structure-determination software program, such as Xplor-NIH or CYANA. After some heavy computations, poof! Out pops a bunch of three-dimensional structures of calmodulin (Fig. 3.13). But why are there multiple structures? Which one is right? And, how did the computer calculate these structures in the first place?

Let’s start with the last question: how does the computer program create protein structures from merely a list of NOE restraints. There are several approaches, but one method implemented in Xplor-NIH is known as simulated annealing. This process is a bit like cooking dry spaghetti. You start off with a stiff rod of pasta and throw it into a boiling hot water (Fig. 3.14, left). When the pasta warms ups and adsorbs the water, it loosens up and starts wiggling around. The string of pasta samples many different conformations as it tumbles around in the hot water. If we take the strand of spaghetti out of the water and cool it down far below room temperature, the pasta would get “stuck” in some random conformation (Fig. 3.14, right).

What happens if we repeat the experiment with a fresh strand of spaghetti? The pasta would wiggle randomly in the hot water as before, but when we cool it down the spaghetti would “fall” into a completely different shape than the pasta in the first experiment (Fig. 3.14, right). Why? Because the initial conditions at the moment the pasta hits the water are slightly different in each case, and thus, the end result will be different.

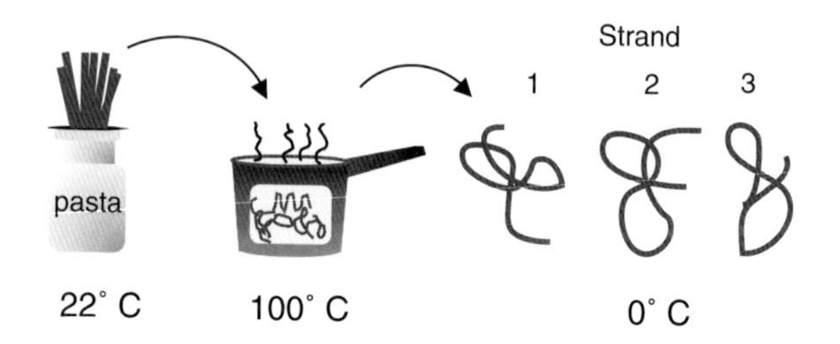

Fig. 3.14 Simulated annealing is similar to cooking pasta: (**a**) Start with stiff rods of pasta; (**b**) throw them into hot, boiling water; and (**c**) when they cool down, they freeze into one configuration

For NMR structure calculations, we can also start with a long, stiff protein rod, like the dried pasta. We take our protein (in this case calmodulin) and adjust all the backbone torsion angles to make an extended conformation (Fig. Fig. 3.15a). For the protein, this wiggling involves rotating torsion angles in both the backbone and the sidechains. At the beginning when the water is very hot, the algorithm allows the protein to move around randomly like a boiling noodle. But as the water begins to slowly cool, the program starts to restrict the protein's wiggling according to the laws of chemistry *and* the NOE distance restraints. Specifically, the software program ensures the following:

(a) All the bond lengths and angles don't stray from their "natural" values, that is the standard bond lengths and angles typically observed in crystal structures of small molecules and proteins.
(b) Atoms don't bump into each other or overlap. Atoms need their space and can't get too close to each other. (That's known as van der Waals interactions.)
(c) The distance restraints provided in the NOE list are satisfied.

If the protein starts folding into a configuration where one of these three requirements is violated, the software program will nudge the protein back into a conformation that preserves these properties.

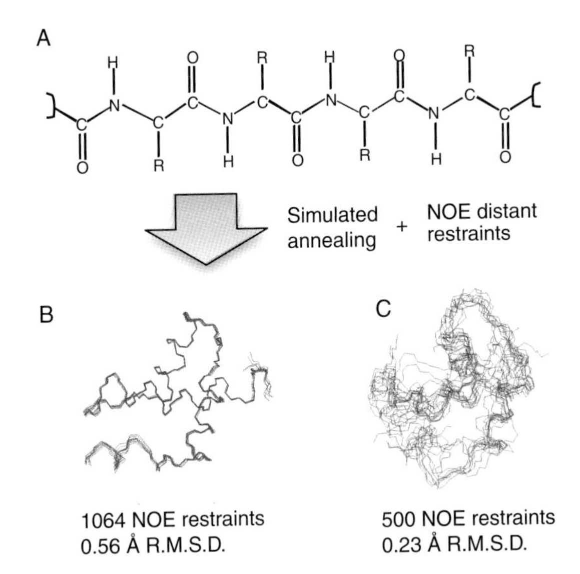

Fig. 3.15 (**a**) In NMR structure determination we start with a protein structure in an extended conformation (**b** and **c**). Then we run simulated annealing many times to fold the protein into a three-dimensional structure that satisfies the laws of protein chemistry and the NOE distant restraints. The more NOE distant restraints we provide the computer, the more similar final structures (ten total structures shown in both **b** and **c**)

To track how well the protein's configuration matches the laws of chemistry and our distance restraints, the computer program uses what's called an "energy function." This isn't real energy, like we've been talking about with NMR, it's merely a way to measure the overall quality of the protein structure. The farther the protein's structure deviates from ideal bond lengths, ideal bond angles, and the experimental distance NOE restraints, the "higher" the structure's energy. The algorithm constantly tries to lower the structure's energy by guiding the protein into a configuration that best matches the three properties above.

What happens if we repeat the calculation with the same extended protein? The random element arises from the fact that the initial velocities for the atoms are assigned random values at the start of the calculation. Will the final structure look like the first one or will it form a completely different conformation like the spaghetti did on its second experiment? This is where those "long-distance" NOE restraints become super important. NOEs between atoms that are far apart in the amino acid sequence tell us what regions of the protein are near each other in the folded state. For example, the cross-peak 137 and H_N and 99-Hγ in Fig. 3.10 tells the program that residue 137 must be cuddled up close to residue 99. These types NOE restraints provide key folding instructions for the software program. If we have enough of these restraints, then the algorithm will guide the protein into the same overall conformation almost every time we perform the structure calculation (Fig. 3.15b). But if we don't have enough distant restraints or many of them are wrong, then the final structure calculated by the algorithm will vary significantly when we run the calculations multiple times (Fig. 3.15b).

Therefore, we keep adding more and more distance restraints to our NOE list until almost all of the calculations create a bundle of structures that have approximately the same fold. We measure the similarity of the structure ensemble by calculating a parameter called the "root-mean-square deviation" or "RMSD." This value merely tells us how close each individual structure is to the average structure in the ensemble. When the RMSD for backbone atoms is <1.0 Å, we know that our NOE data are sufficient to define the fold of the protein and that we are probably on the right track for solving the protein's structure. To be sure, we also check that all the NOE restraints are not violated in this final structure.

The NOE restraints are the main data used to determine structures by NMR, but they are certainly not the only ones. Other types of NMR restraints include torsion angles derived from coupling constant experiments; dipolar coupling restraints that provide orientational information; and chemical shift restraints related to secondary structure. Using these additional types of restraints help improve the precision of the structure calculations.

References and Further Reading

Cantor CR, Schimmel PR, (1980) Biophysical chemistry part I: the conformation of biological macromolecules, Chaps. 1, 2, and 5. W. H. Freeman and Company, New York.

Cavanagh J, Fairbrother WJ, Palmer AG III, Rance M, Skeleton NJ (2007) Protein NMR spectroscopy: principles and practice, 2nd edn., Chaps. 5, 7, and 10. Academic Press, Amerstdam.

Gunter P (1997) Calculating protein structure from NMR data. Methods Mol Biol 60:157–194.

Herrmann T, Güntert P, Wüthrich K (2002) Protein NMR structure determination with automated NOE assignment using the new software CANDID and the torsion angle dynamics algorithm DYANA". J Mol Biol 319:209–227.

Levitt M (2001) Spin dynamics: basics of nuclear magnetic resonance, Chaps. 7 and 16. John Wiley & Sons, Inc., Chichester.

Overhauser A (1996) In: Grant DM, Harris RK (eds) Encyclopedia of nuclear magnetic resonance, vol. I. John Wiley & Sons, Inc., New York, p 513.

Wüthrich K (1986) NMR of proteins and nucleic acids, Chaps. 6–10. John Wiley & Sons, Inc., Chichester.

Chapter 4
Silencing of the Bells: Relaxation Theory Part One

You know, what these people do is really very clever. They put little spies into the molecules and send radio signals to them, and they have to radio back what they are seeing.

(Felix Bloch recalling Niels Bohr's description of nuclear magnetic resonance).
J. Mattson and M. Simon, The Pioneers of NMR and Magnetic Resonance In Medicine: The Story of MRI (Bar-Ilan University Press, Jericho, 1996).

Ahh, relaxation. Say it slowly, and the word itself conjures calming images—a hammock in the sun, a cool breeze at the beach, or a mug of hot chocolate on a snowy day. No matter how you like to wind down after a tough week or stressful semester, relaxation is about releasing pent up energy and returning to a lower energy state that you think of as normal.

But to reach a state of zen takes time. You don't finish a three-hour exam, and then instantaneously return to your calm, cool self. Most of us need at least a few days to release pent up energy and completely relax.

A great way to accelerate the relaxation process is to exercise. Go for a run, hit the gym, even go dancing! Exercise is a second type of

M. Doucleff et al., *Pocket Guide to Biomolecular NMR*,
DOI 10.1007/978-3-642-16251-0_4, © Springer-Verlag Berlin Heidelberg 2011

relaxation that we humans need. It's full of chaos and disorder, but it frees and defocuses the mind just like the calmer type of relaxation.

Like stressed out humans, ringing nuclei also have extra energy. And, they have multiple ways to release the excess energy, defocus their signal, and relax back to the low-energy state. No matter the route, the result is the same—the nuclei return to where they started before the NMR experiment, and our precious NMR signal vanishes.

In many cases, this relaxation is a nuisance because it leaves us with less than ideal NMR data. But for large macromolecules, like proteins and nucleic acids, the relaxation is a blessing in disguise. Hidden inside these diminishing electromagnetic waves are secrets about the dynamics and motion of the molecules, information that's nearly impossible to obtain by any other biochemical or biophysical technique.

So turn on the Norah Jones, grab a cool beverage, and get ready to learn about how atoms kick back and *relax*.

4.1 Nothing Rings Forever: Two Paths to Relax

The simplest NMR experiment involves two steps (Fig. 4.1a):

(a) Ring all hydrogen atoms.
(b) Collect the electromagnetic waves they create.

Unfortunately, the sinusoidal wave produced by an excited nucleus doesn't last forever. The amplitude of the wave slowly decreases until the signal withers into a straight line (Fig. 4.1b). This signal decay is called *relaxation*, and it's the focus of the rest of the book.

As we humans enjoy different types of relaxation to return to "normal" after a stressful event, nuclei also have two major routes for relaxing back to the resting state. Even without knowing an iota about quantum mechanics or "relaxation theory," we can predict the two different mechanisms causing the decay in signal.

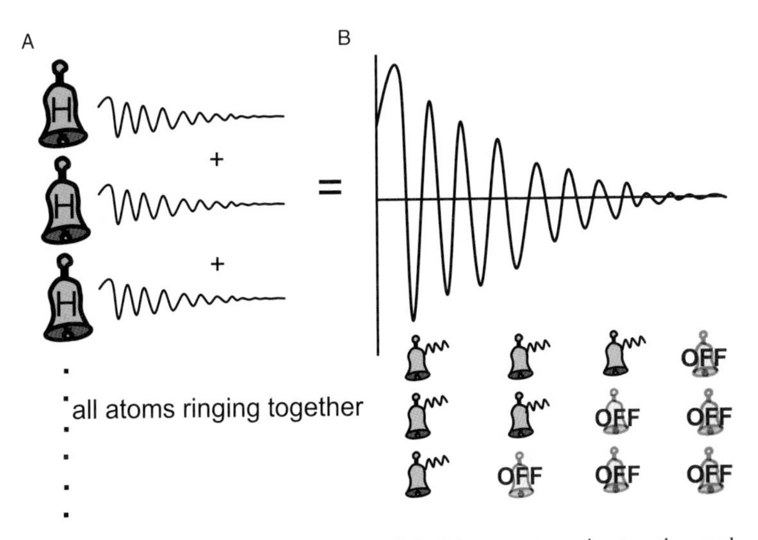

Fig. 4.1 (**a**) In an NMR sample, many nuclei of the same type ring together, and their electromagnetic waves combine to create the NMR signal that we detect. (**b**) As the bells stop ringing, or "shut off," the NMR signal decays and eventually vanishes

Imagine a choir of a million identical bells, all ringing the exact same note. The bells are very tiny so each one only whispers the note (Fig. 4.1a), but because there are so many bells and they all start playing at the same time, their waves combine together and the note is easily detected. This note is our NMR signal (Fig. 4.1b), and the bell choir is a specific type of atom in the NMR sample, such as the amide proton of residue 137 in calmodulin. [Remember, the NMR sample contains 10^{13} molecules, so we have approximately 10^{13} 137-H_N atoms "singing" a note at ~500 MHz or specifically 9.104 ppm] (to learn more about how many atoms are "singing," see Mathematical Sidebar 4.1).

So what stops the choir? What causes the note at 9.104 ppm to go from a gentle forte to an imperceptible pianissimo? It's obvious, right? Individual bells stop ringing!

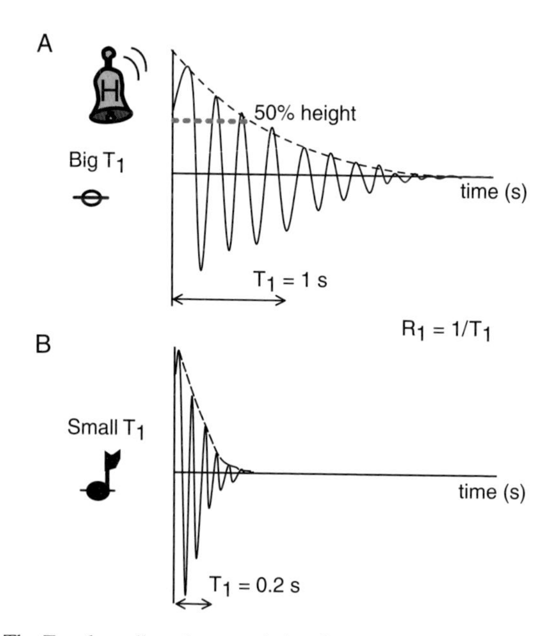

Fig. 4.2 The T_1 value tells us how much time it takes for the nuclei to stop ringing. (**a**) Atoms with a long T_1 value ring for a long time, like a whole note; (**b**) atoms with a short T_1 value turn off quickly, like an eighth note. Specifically, it requires ~$T_1 \times 0.7$ seconds for half the bells to turn off (gray dotted line in **a**)

A ringing nucleus can spontaneously shut off just like a light switch. Every nucleus contributes to the NMR signal (Fig. 4.1a), so when they start turning off, the amplitude of the electromagnetic wave decreases. When all the nuclei are "off," the NMR signal vanishes (Fig. 4.1b).

This mechanism of relaxation is called "T_1," and the "T_1 value" tells us the time it takes for the high-energy nuclei to stop ringing and return to their low-energy state. Atoms with large T_1 values can play long, drawn out notes that last more than a second (Fig. 4.2a). But atoms with short T_1 values play short eighth notes that last only a fraction of a second (Fig. 4.2b).

Another way to think of T_1 is that it takes approximately "($T_1 \times 0.7$) seconds" for half the nuclei to stop ringing (dotted line in Fig. 4.2a) ($1/T_1$ is the exponential rate of decay due to T_1 relaxation). Sometimes spectroscopists talk about T_1 relaxation in terms of a rate, R_1, which is simply $1/T_1$. R_1 tells us how quickly the bells shut off. Just remember "T" is for the *time* it takes for the bells to stop ringing and "R" is for the rate at which the bells become silent. We'll learn about the details of T_1 relaxation in a few sections, but first, let's examine the second relaxation pathway for ringing nuclei.

Okay, so the amplitude of the NMR signal decreases as individual nuclei in the sample stop ringing. That makes good sense, but can you think of another mechanism that would decrease the intensity of the electromagnetic wave? Here's a hint: ringing nuclei should probably invest in a tuning fork.

Each 137-H_N atom in the NMR sample creates a tiny electromagnetic wave. All 10^{13} of these waves sum together to create the NMR signal that we detect in the spectrometer. When these waves all have exactly the same frequency and start ringing together, they amplify each other, producing one unified wave that is much stronger than an individual wave (Fig. 4.3a).

But what happens if the atomic bell choir can't stay "in tune"? If the 137-H_N hydrogen atoms begin playing notes at slightly different frequencies, the individual waves start to "fall out of sync," or *out of phase*, with each other—some waves go up while others go down (Fig. 4.3b). Instead of reinforcing or augmenting each other, the waves begin to cancel each other out! This is called *dephasing*, and it diminishes the NMR signal because the positive parts of the waves equal the negative parts of the wave and they sum to zero (Fig. 4.3b).

Dephasing underlies the second relaxation mechanism for ringing atoms, called "T_2." The T_2 value tells you how quickly the individual electromagnetic waves become "out of sync" due to fluctuations in their ringing frequencies. Hydrogens with large T_2 values can hold the same name note for an extended period of time (Fig. 4.3a) whereas atoms with small T_2 values quickly change their frequency, causing rapid loss of signal (Fig. 4.3b). T_2 is also called the *transverse relaxation* time and *spin-spin relaxation* time (Table 4.1).

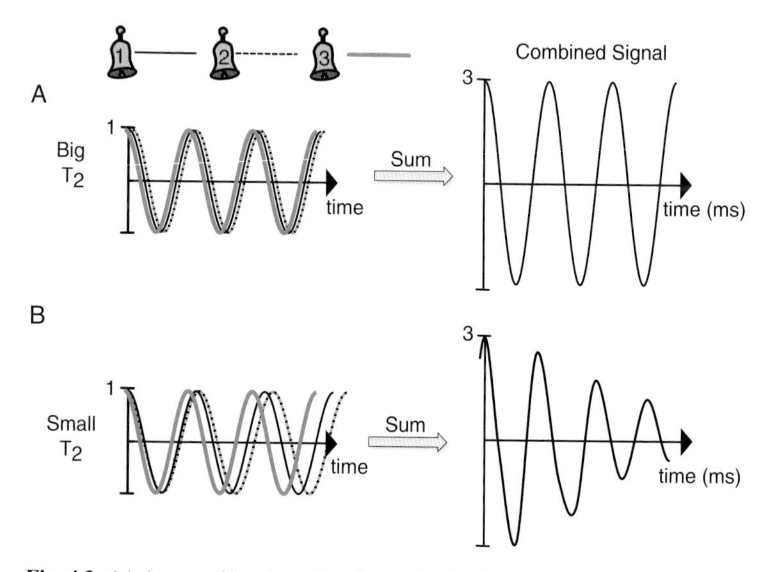

Fig. 4.3 (**a**) Atoms with a long T_2 value maintain the same ringing frequency for a long time, leading to an NMR signal that stays strong over time. (**b**) In contrast, atoms with a short T_2 value quickly fall "out-of-tune" or shift their ringing frequency, leading to a short NMR signal that decays quickly

Table 4.1 Two types of relaxation

Relaxation type	Easy description	More detailed description	Alternate names
T_1	The time it takes for the atomic bells to "turn off"	The time it takes for the nuclei to return their magnetic needles to the *z*-axis.	1. Longitudinal relaxation 2. Spin-lattice relaxation
T_2	The time it takes for the atomic bells to fall out of tune.	The time it takes for NMR the signal to decay due changes in the ringing frequency (dephasing)	1. Transverse relaxation 2. Spin-spin relaxation

As with T_1, it takes ~$T_2 \times 0.7$ for the signal to drop in half due to dephasing. Again, this is because $1/T_2$ is the exponential rate of signal decay due to T_2 relaxation. Also like with T_1, T_2 is often talked about as R_2, which is simply $1/T_2$. R_2 tells us the rate of dephasing.

We'll talk more about the dephasing and T_2 in Sect. 4.4, but first, why do we care about how the NMR signal decays? What's interesting about that? What's the big deal with this atomic relaxation?

4.2 Relaxation: Ticket to the Protein Prom

Relaxation is our ticket to the biomolecular dance. The decaying NMR signal possesses surprisingly rich information about the internal motion of macromolecules and provides a rare glimpse of how biomolecules function during catalysis and signaling.

But why? How does the decay in the NMR signal (Fig. 4.1) provide information on the dynamics of proteins and nucleic acids? To answer this, we need to go back to Chap. 1 and learn a bit more about what makes the atoms ring at a particular frequency in the first place. In other words, what exactly is that "big hammer" at the beginning of our NMR experiments.

In Chap. 1 we learned that an atom's ringing frequency depends directly on the strength of the surrounding magnetic field. In an 11.8-Tesla magnet, the hydrogen atoms ring at approximately 500 MHz, and at 18.8 Tesla, they ring at approximately 800 MHz (Fig. 4.4). However, the *exact* ringing frequency (or chemical shift) of a specific hydrogen depends not only on the external field but also on local magnetic fields created by the surrounding electrons and other atoms (remember that a magnetic field is created by moving electrons). Therefore, if a portion of a protein, such as a loop or domain, changes conformation, the local electromagnetic field surrounding the particular atom will also change, causing the atom's ringing frequency to shift slightly.

For example, the H_N atom in Fig. 4.5a is initially like most amide hydrogens in a protein and rings at about ~8.5 ppm. But when the

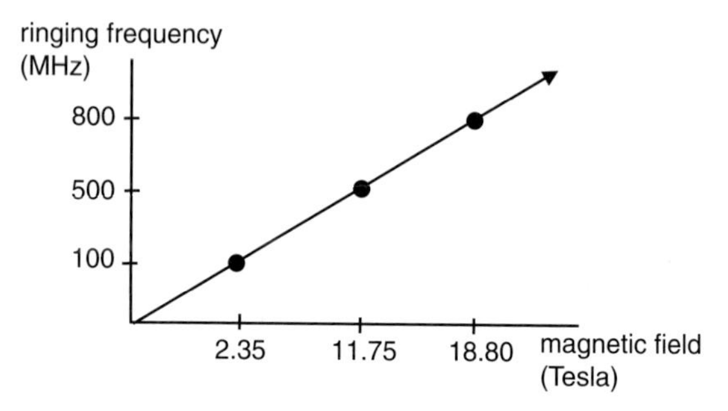

Fig. 4.4 A nucleus"s ringing frequency (also called the Larmor frequency) is linear with the strength of the magnetic field

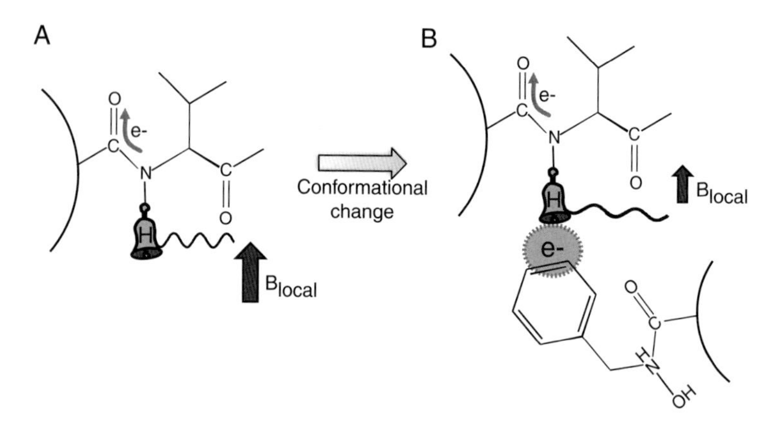

Fig. 4.5 (**a**) In the initial protein configuration, the peptide bond strips the electrons away from the amide hydrogen (shown as a bell), creating a strong local magnetic field (B_{local}) and a high ringing frequency. (**b**) A conformational change brings an aromatic ring near the amide hydrogen, decreasing the strength of the local magnetic field (B_{local}) and dramatically lowering the ringing frequency of the amide hydrogen

protein alters its conformation, the hydrogen's electromagnetic environment changes drastically—the amide hydrogen now sits right on top of π-electrons of an aromatic ring (Fig. 4.5b). These electrons significantly reduce the total magnetic field felt by the H_N, causing its ringing frequency to shift almost 1 ppm or 500 Hz (at 11.8 Tesla)!

As we just learned, altering the ringing frequency speeds up T_2 relaxation, making our signal decay faster (Fig. 4.3b). So, there we have it—a connection between relaxation and the dynamics of macromolecules!

There's no harm in thinking about relaxation and motion in terms of this simple idea, but this is merely the beginning of relaxation's complicated and sometimes twisted relationship with molecular dynamics. To understand the full story, we need to learn more of the details of our NMR experiments, especially about what makes an atom ring in the first place.

Mathematical Sidebar 4.1: Boltzmann Distribution

When a sample is placed into an NMR spectrometer the individual spins align with the magnet field. For protons, ^{13}C, ^{15}N, and other atoms with a nucleus of spin ½, there are two possible ways for these spins to align: pointing up with the magnetic field or pointing down against the field (Fig. M4.1). Nuclei aligned with the field are in a lower energy state than those pointing in the opposite direction of the magnetic field. This leads to two distinct states with a difference in energy given by

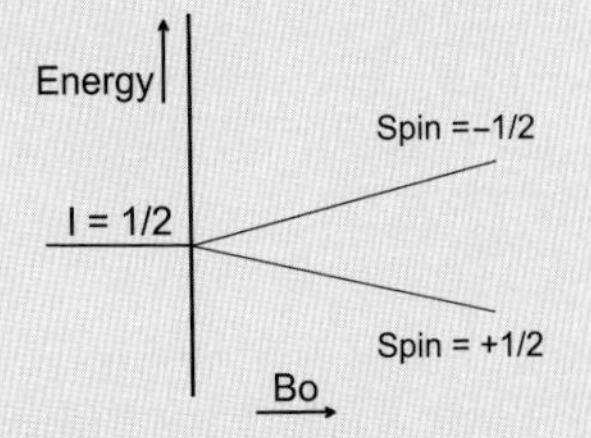

Fig. M4.1 The energy difference for spins (I) aligned with and opposing the external magnetic field B_o

$$\Delta E = \frac{h\gamma B_o}{2\pi}$$

where γ is the gyromagnetic ratio, B_0 is the strength of the magnetic field, and h is Planck's constant. Note that as B_0 increases, so does the energy separation. This explains why the resolution of ringing frequency increases as the strength of the magnetic field increases.

How many spins populate each state? Fortunately for us, the two states are populated unequally, and this population difference is what gives rise to the intensity of our NMR signal. That is, the greater the population difference, the stronger our signal. The population difference is dependent upon both the energy separation of the two states (ΔE) and the temperature of the system (T). Specifically, the numerical population difference is given by the Boltmann distribution;

$$\frac{N_\alpha}{N_\beta} = \exp\left(-\frac{\Delta E}{kT}\right)$$

where k is the Boltzmann constant and N_α and N_β are the number atoms in the lower and higher energy states, respectively. Thus, the larger the energy separation, the larger the population difference (N_α/N_β) and the stronger the signal. The energy separation for protons is significantly greater than that for carbon atoms (recall that the γ of ^{1}H is four-times that of ^{13}C). This explains why one can quickly collect a spectrum when we detect ^{1}H atoms, but we need far more scans (and thus increased signal averaging) to get the same intensity when we detect the ringing of ^{13}C atoms.

4.3 Oh-My, How Your Field Fluctuates

Have you ever used a compass? Probably not, given the abundance of cheap GPS devices now available. But back in the olden days, like

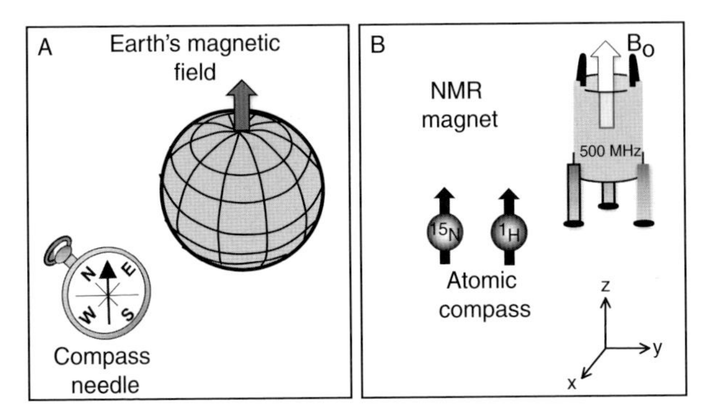

Fig. 4.6 (**a**) The magnetic needle in a compass aligns with the magnetic field of the Earth. (**b**) Analogously, the tiny little magnets in the ^{1}H, ^{15}N, and ^{13}C nuclei align with the super-strong magnetic field (B_0) pointing toward the ceiling inside the NMR probe

in the 1980s, the compass was an essential navigation device while exploring the wilderness or sailing on the ocean.

Imagine sitting on a sailboat—ocean to the left, ocean to the right, ocean in the front and behind—no direction to be found. Now place a compass in your hand. The tiny little bar magnet wobbles around, and then settles itself—north! The compass identifies the North Pole for you by aligning with the Earth's magnetic field (Fig. 4.6a).

The nuclei of particular atoms, such as ^{1}H, ^{15}N, and ^{13}C, contain tiny, "invisible" magnets that are quite similar to the bar magnet in the compass—they like to align with magnetic fields. In an NMR spectrometer, we have a super-strong magnetic field pointing toward the ceiling (or along the z-axis). So when we put the protein or DNA sample in the magnet, the hydrogen nuclei line up vertically like toothpicks in a box (Fig. 4.6b).

This is how all NMR experiments begin—hydrogen nuclei standing up straight along the z-axis in response to an external magnetic field (Fig. 4.7a) (to learn why this happens, see Mathematical Sidebar 4.1). Now what happens when we apply a magnetic field perpendicular to the z-axis, say along the x-axis? Like miniature compasses, the nuclei

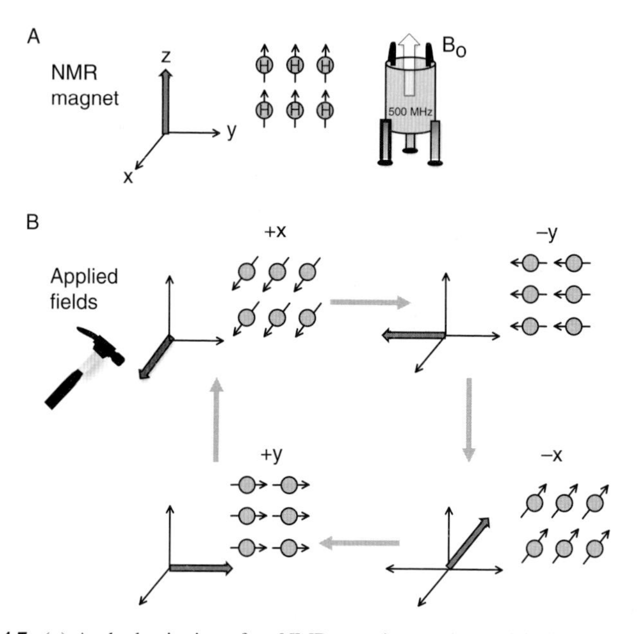

Fig. 4.7 (**a**) At the beginning of an NMR experiment, the nuclei align their magnetic needles with the external field (B_o) pointing along the +*z*-axis. (**b**) Then we apply a strong electromagnetic field in the *xy*-plane, which causes the nuclei to rotate around the *z*-axis, pointing their magnetic needles at: +*x* (top left), -*y* (top right), +*y* (bottom right), −*x* (bottom left), and then back at +*x* (top left) to repeat the cycle

wobble around and eventually align with this new direction of the magnetic field (Fig. 4.7b)

Now imagine rotating this magnetic field in circles around the *z*-axis like a record turntable. First the applied magnet points along the +*x*-axis (Fig. 4.7b); then it moves to −*y* and then to −*x*; next it's along the +*y*-axis, and then finally it starts all over again at the *x*-axis. During this time, the little compasses in the nuclei try to keep up and align with the rotating magnetic field. As a result, they start spinning around the *z*-axis, like they are sitting on top of a record player (Fig. 4.8)

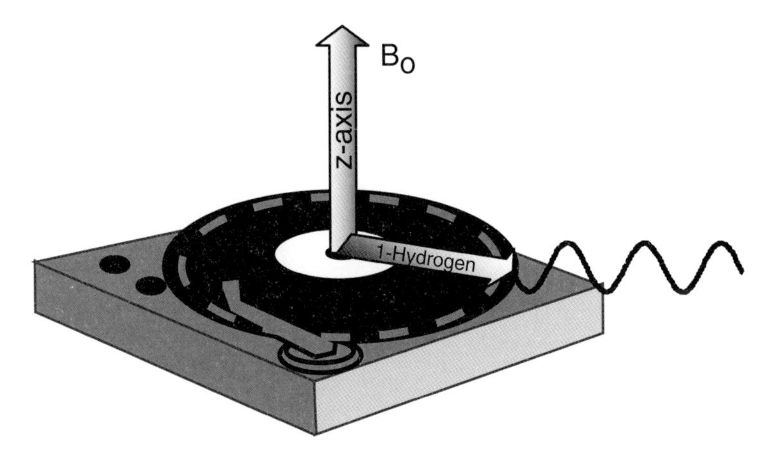

Fig. 4.8 The electromagnetic wave applied in the *xy*-plane (the "hammer") forces the nuclei to *spin* or rotate around the *z*-axis, like a record on an old-school turntable

When a magnetic field is changing directions or oscillating, what does it create? You got it—an electromagnetic field! As the nuclei revolve around the *z*-axis, they create a fluctuating magnetic field. This is how the nuclei ring and create our NMR signal!

Ok, this all seems reasonable: each nucleus contains a tiny compass that aligns with a magnetic field. When that field starts spinning in the *xy*-plane, the tiny bar magnets in the nuclei align with the revolving field and start spinning too. But there's one problem here. The original field in the *z*-axis is incredibly strong, one of the strongest magnets in the world. To flip the nuclei's magnet down into the *xy*-plane (Fig. 4.7b), do we need to apply a field that's as powerful or even more powerful than this magnetic field pointing toward the ceiling?

This is where NMR becomes amazing, almost magical. Instead of using brute force to flip the nuclei's magnets around the *z*-axis, we use an elegant trick called *resonance*. A remarkable phenomenon, resonance, allows relatively weak forces to have huge effects. The key to

Fig. 4.9 Pushing an NFL lineman on a swing is easy if you take advantage of the phenomenon called resonance and give him many small pushes at the apex of his motion

resonance is timing; the weak forces must be "in sync" with the natural rhythm or frequency of the system, so that every iota of energy builds up over time.

Think of pushing a 500-lb NFL lineman on a playground swing (Fig. 4.9). To start him swinging, you could use all your strength and apply a few tremendously powerful pushes. Or, you could be more elegant and simply give the behemoth many small pushes every time the swing returns to you (Fig. 4.9). Although the pushes are relatively weak, their effect accumulates over time because their timing matches the natural swinging frequency. With resonance to assist you, the huge man is flying high without much effect at all!

Like the swinging lineman, the tiny magnets in nuclei also have a natural frequency at which they "ring," or rotate around the z-axis (Fig. 4.7). If we apply an oscillating electromagnetic field in the xy-plane with a frequency that nearly matches this preferred spinning frequency, even a relatively weak field will eventually flip the magnets down into the xy-plane and start them spinning around the z-axis

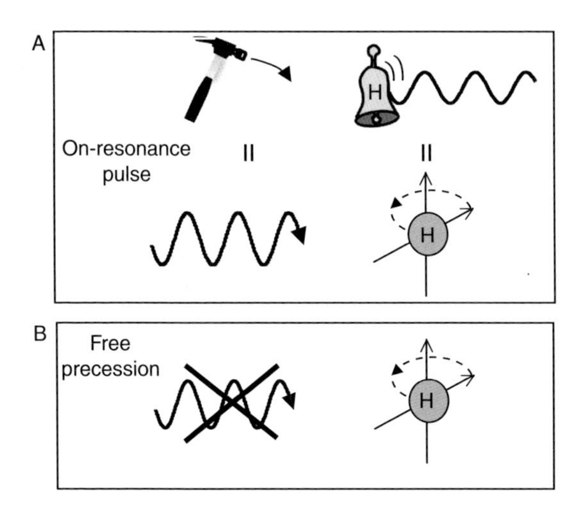

Fig. 4.10 (**a**) In an NMR experiment, the "hammer" that forces the nucleus to spin or rotate around the z-axis is actually an electromagnetic wave in the xy-plane set to the natural rotating frequency of the nucleus. (**b**) When this electromagnetic wave or pulse is turned off, the atoms continue to spin in the xy-plane at their natural frequency in *free precession*

(Figs. 4.7b and 4.8). The applied electromagnetic field "keeps up" with nuclei, continuously giving them little nudges as they rotate in the xy-plane.

The cool part is that once we shut off this fluctuating electromagnetic field, the nuclei continue to spin at their favored frequency (Fig. 4.10b), just as the lineman's swing continues to oscillate at the swing's natural frequency when you stop pushing him. This "free" rotation around the z-axis at the resonant frequency is called *free precession* or *evolution*.

Now the details of the NMR experiment become clearer. That big hammer at the beginning of the experiment is actually a fluctuating electromagnetic wave (Fig. 4.10a) in the xy-plane with a frequency matching the natural ringing frequency of the nucleus. And, the "ringing" that we detect at the end of the experiment is an electromagnetic wave created by a tiny magnet inside the nucleus rotating around the z-axis (Fig. 4.10b).

This last part is the key idea—weak magnetic fields can a have a significant effect on nuclei when the electromagnetic field oscillates at a frequency close to their resonant frequency. With this idea in mind, the connection between relaxation (or the decrease in signal decay) and molecular motion is obvious:

Molecular motions produce oscillating magnetic fields that interfere or disrupt the resonance ringing in our NMR signal. Repeat the idea if you need to because it is critical for studying the dynamics of proteins. Let's see how it works.

As we just learned, each hydrogen in a protein has a tiny magnet inside its nucleus (Fig. 4.11a). When the nucleus changes locations—whether it's moving in a flexible loop or tumbling with the rest of the protein in a solution—it creates a fluctuating magnetic field nearby,

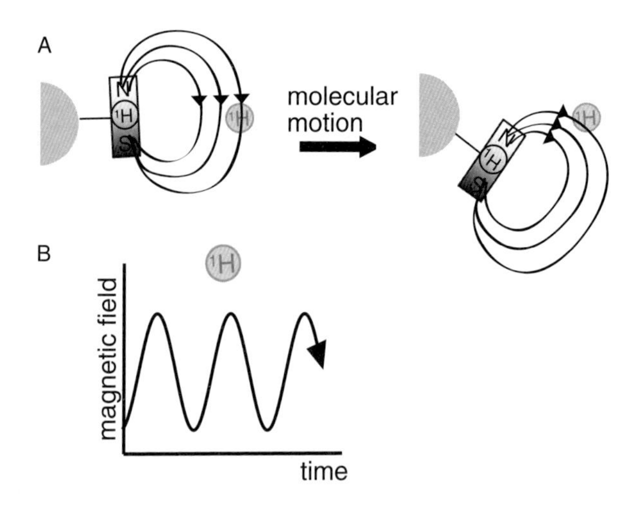

Fig. 4.11 (**a**) The hydrogen nucleus produces a magnetic field, which changes direction and strength as the molecule moves through space (**b**) Thus, the magnetic field felt by nearby hydrogen atoms (gray circle) fluctuates over time

which changes the direction and magnitude of the local magnetic field over time (Fig. 4.11b). Yes, this field is weak and affects only hydrogens in its own neighborhood. However, when the motion has a frequency close to the resonance frequency of a nearby nucleus, then BAM! You guessed it—these relatively weak magnetic fields have a significant impact on the ringing of nearby nuclei.

What about all those electrons whizzing around in covalent bonds? We already know that moving electrons produce magnetic fields too. So when these bonds tumble around in solution with the rest of the protein or move during structural rearrangements, they also create fluctuating magnetic fields that can affect nearby nuclei too.

The problem with all these local oscillating magnetic fields is that they are never "in sync" with the fluctuating field that we apply at the beginning of our NMR experiment. Our field provides precise instructions for the spinning nuclei: go to the +*x*-axis, now to the +*y*-axis, then to the -*x*-axis, etc. All the nuclei follow the directions and rotate around the *z*-axis together (Fig. 4.12a).

In contrast, the neighboring nuclei and electrons create oscillating fields that sound like random gibberish in comparison to the precise commands of our the applied pulse. These fluctuating fields tell the nuclei to go in all different directions—45° to the southwest, 60° to northeast, et cetera. These random commands disrupt and counteract the orderly ringing established by our applied electromagnetic field. Instead of rotating around the *z*-axis "in sync," each nucleus begins charting its own coarse. This causes their electromagnetic waves in the *xy*-plane to start canceling each other out, and our NMR signal decreases (Fig. 4.12b).

So to understand the internal dynamics creating these local oscillating fields, we need to characterize how the NMR signal decays. This is where T_1 and T_2 come to help us. Each type of relaxation reports on different kinds of motion inside a molecule. (Think of them as tiny newscasters stationed inside the nucleus, transmitting information about the fluctuating fields around them by diminishing the NMR signal.) Let's learn more about these high-tech correspondents.

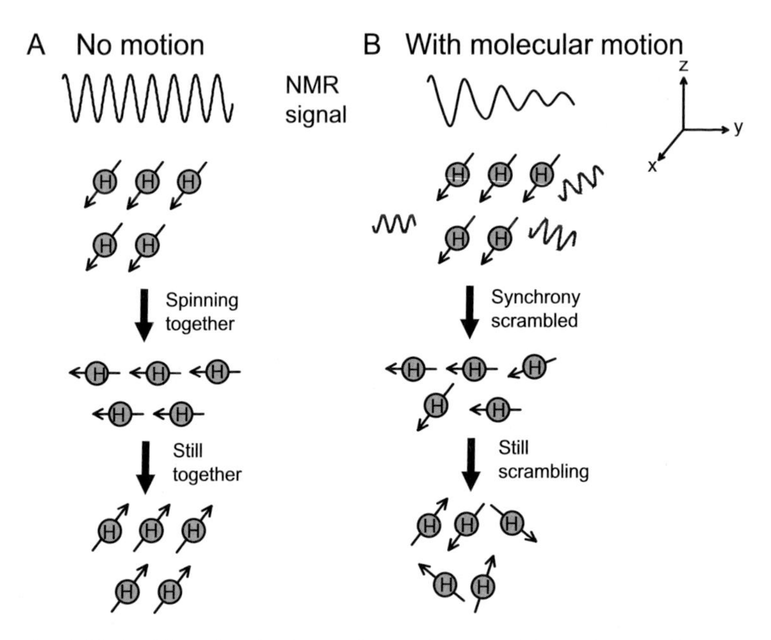

Fig. 4.12 (**a**) Without molecular motion, hydrogen nuclei stay together "in synchrony" as they rotate in free precession around the z-axis, producing a strong NMR signal. (**b**) In contrast, molecular motion creates random fluctuating magnetic fields that disrupt the synchronous free precession and scrambles our NMR signal

4.4 Blowing Off Steam and Returning to Equilibrium: T_1

Have you ever played tetherball? Perhaps you've seen this funny little game at family picnics: there's a pole stuck in the ground, a rope tied at the top, and a volleyball at the end of the rope (Fig. 4.13a). You hit the ball to your right, and the ball begins to circle around the pole at a specific frequency counter-clockwise. If you give the ball a nudge to the right every time it passes by you, the angle between the ball

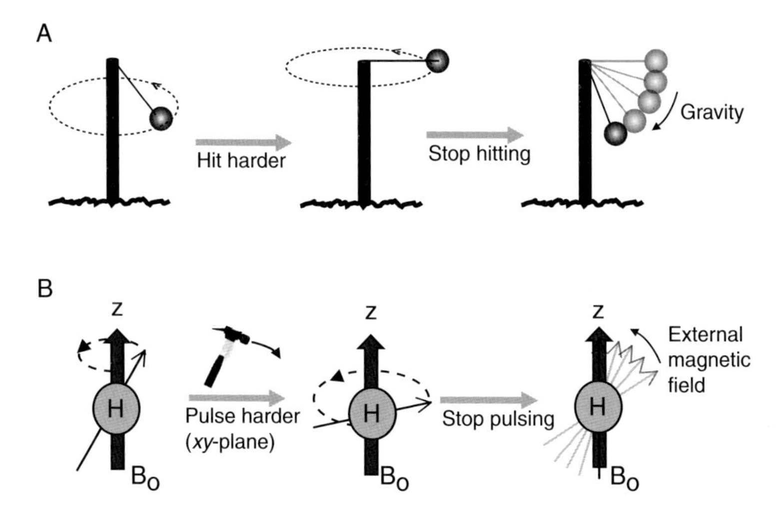

Fig. 4.13 (**a**) Hitting a tetherball makes it spin around the pole (left) and drives the ball into the horizontal plane (middle). When you stop hitting it, the ball falls back to the pole in response to gravity (right). (**b**) In NMR, an electromagnetic pulse makes the hydrogen's nucleus rotate around the *z*-axis (left) and drives it into the *xy*-plane (middle). When the pulse is turned off, the nucleus returns to the +*z*-axis in response to the strong magnetic field pointing upward (right)

and the pole will slowly grow (Fig. 4.13a, left) until the rope is completely horizontal in the *xy*-plane and 90° from the pole (Fig. 4.13a, middle).

What happens when you stop hitting the volleyball? It will continue to rotate around the pole because the ball has angular momentum. But the angle between the ball and the pole will slowly decrease as gravity pulls the volleyball back down to the ground (Fig. 4.13, right). The energy that you put into the system by hitting the ball is lost as the ball falls down to the pole.

An NMR experiment is quite similar to this tetherball game except that it's "upside-down" (Fig. 4.13b). We "hit" the nuclei with an electromagnetic wave in the *xy*-plane (Fig. 4.13b, left), and the little magnets spin around the *z*-axis, eventually reaching the *xy*-plane (Fig. 4.13b, middle), just as the tetherball reached the top of the pole when you smack the ball hard (Fig. 4.13a, middle). When we shut off the fluctuating field in *xy*-plane (stop the "hammer"), the nucleus's magnet keeps rotating around because the nucleus has angular momentum, just like the swinging volleyball (Fig. 4.13a).

But nuclei don't care about the gravitational force pointing downwards. Instead, they respond to the huge magnetic force (11.75 Teslas) pointing up toward the ceiling (the *z*-axis). So instead of slowly "falling" back down to the ground because of gravity, the nuclei gradually turn their magnets upwards in alignment with the strong external magnetic field (Fig. 4.13b, right). The extra energy given to the nuclei by the applied electromagnetic wave (the "big hammer") is lost to the surroundings as they return home to the *z*-axis.

This is T_1 relaxation—the nuclei releasing their extra energy as they return their little compass needles back in alignment with the large magnetic field along the *z*-axis (see Mathematical Sidebar 4.2 for more on T_1 relaxation).

Why is this important? For starters, the T_1 relaxation rate tells us how long we need to wait between NMR experiments. All NMR experiments start with nuclei pointing their magnets up towards the +*z*-axis. So, before beginning another experiment, we need to wait for the nuclei to do just that—return their magnets back to +*z*-axis at the end of first experiment (Fig. 4.13b, right). This is called the *recycle delay* and is optimally set to approximately five times the T_1 relaxation rate (~1–3 seconds for proteins and DNA molecules).

Remember that the signal we detect in an NMR experiment is a *fluctuating magnetic field in the xy-plane only* (Fig. 4.13b, middle). Small magnetic fluctuations along the *z*-axis are masked by the gigantic external magnetic field pointing along +z (11.75 Tesla); therefore, we don't even bother looking there. We just point our electromagnetic radar at the *xy*-plane and disregard fluctuations along the *z*-axis. Once the nuclei's magnets return straight up along the *z*-axis (Fig. 4.13b, right), their magnetic field in the *xy*-plane vanishes and so does our

NMR signal. T_1 tells us how fast this occurs: *nuclei that like to hang out in the xy-plane have long* T_1 *values* (Fig. 4.2a), and *nuclei that prefer to return back home to the z-axis have short* T_1 *values* (Fig. 4.2b).

Here are few other key ideas to remember about T_1 relaxation:

(a) T_1 relaxation has two alternate names:

a. *Spin-lattice relaxation* because the spinning nuclei release their extra energy to the "lattice" or "surroundings" as they return to equilibrium (aligned along the *z*-axis).[1]

b. *Longitudinal* relaxation because it tells us how fast the nuclei return back to the "longitudinal" or "*z*" axis (Table 4.1)

(b) T_1 depends greatly on the size of the molecule. A plot of T_1 versus molecular weight is shaped like a "U," with medium-size

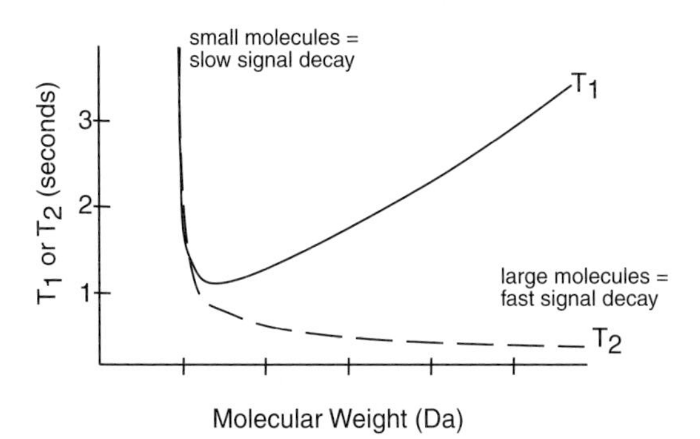

Fig. 4.14 Graph of T_1 (solid line) and T_2 (dotted line) versus molecular weight of a molecule. For large molecules, like proteins, the T_1 value is significantly larger than the T_2, so the T_2 determines the lifetime of our NMR signal

---

[1]Spin-lattice relaxation is a confusing name because energy is released to the surroundings in both types of relaxation, T_1 and T_2 but that's the way it is!

molecules relaxing the fastest and having the smallest T_1 values (Fig. 4.14).[2]

(c) T_1 is always greater than (or equal to) T_2 (Fig. 4.14). Therefore, it is T_2, not T_1, which limits the time we have to acquire the NMR signal. This means that it is T_2, not T_1, which also determines the width of the peak in the NMR spectrum. Let's see why this is true.

What is your T_1 value—do you like to stay for a long time in a high-energy state or do you hurry home after a stressful day of work?

Mathematical Sidebar 4.2: T_1 Relaxation

T_1 measures the time it takes for the nuclei to return to the "longitudinal" or z-axis. At the start of an experiment, the nuclei are "lined up" with the magnetic field (along the z-axis), which results in a net magnetization I_z. This is also called the equilibrium magnetization I_o because at equilibrium, the spins are aligned with the field. When we "hit" these nuclei with an electromagnetic field, it moves the nuclei into the xy-plane, resulting in a net magnetization along the x-axis I_x. When the applied electromagnetic field is turned off, the spins will relax back to equilibrium (i.e., re-align with the field) by longitudinal relaxation T_1. The rate of the return to the z-axis is given by a first-order rate expression:

$$\frac{d}{dt}I = \frac{-(I - I_o)}{T_1}$$

where I is the signal intensity (along the x-axis or y-axis) at a particular time t and I_o is the equilibrium value described

[2] To learn why this curve is shaped like a "U," see Chap. 16 of "Spin Dynamics" by M. Levitt (2001).

above. (Note: we can only "see" magnetization in the *xy*-plane, thus I_o does not contribute to our NMR signal.) If you measure the signal right after the applied electromagnetic pulse is turned off, the intensity along the *x*-axis will be large. But if you wait a period time *t*, a subset of the nuclei will have relaxed back to equilibrium (i.e., aligned with the field), and the intensity in the *xy*-plane will decrease. Thus, by measuring the intensity *I* at a number of time points after the pulse and integrating the equation above, you can determine the longitudinal relaxation time T_1.

4.5 Loosing Lock-Step[3]: Coherence and T_2

Okay, so T_1 tells us how quickly the nuclei return their little compass needles to the z-axis, destroying our precious signal in the *xy*-plane. But what about T_2? How does it contribute to decay in NMR signal? To answer this question, we need to learn more about a concept called coherence.

When you hear the word "coherent," you probably think about whether something "makes sense" or is "comprehensible." Perhaps the word even conjures up memories of times when you were "incoherent"—like when you're trying to explain last week's physics lecture to your friends. Yes, in the "real world," coherence means *logical*, *intelligible*, and *consistent*. But for NMR spectroscopists and other physical chemists, coherence has a totally different meaning: it's all about synchrony and coordination of electromagnetic waves.

Think of watching a college marching band from the top bleachers of a stadium. Although the band contains hundreds of musicians,

---

[3]Comparing phase coherence to walking in lock-step is the idea of Nobel Prize-winning physicist Wolfgang Ketterle. He used this analogy in 1995 when he and his laboratory created one of the first Bose-Einstein condensates—a new state of matter in which atoms line up in lock-step at super low temperatures to become a *coherent* wave.

Fig. 4.15 Marching lockstep is a beautiful example of coherence—at any moment in time, the angle between the left and right legs (θ) is exactly the same for every member of the marching band

the students appear to move as one unit on the field, like a giant millipede. The cohesion arises from the marching synchronization: At any moment in time, the angle between the left leg and right leg will be exactly the same for each member of the band (Fig. 4.15). To maintain this *lockstep marching* requires two conditions: First, the band members must start marching together. Second, the musicians must walk at the same pace. A slight change in stride frequency amongst the band members will quickly scramble their stepping synchrony and diminish their fluid, cohesive motion.

Electromagnetic waves can also walk in *lockstep*, and when they do, physicists say that the waves are *coherent.*

Like the band members, waves must maintain the same pace or frequency. But they also must start off at the same point in the wave. Physicists call the starting point of a wave, its "phase." Table 4.2 shows how it works. Waves that begin at $y = 1$ have a phase equal to 0; those starting at -1 have a phase of 180° (or π); and, waves starting

Table 4.2 The rules for phase angles

Position at time = 0	Phase angle (°)	Phase angle (radians)
1	0	0
−1	180	π
0	90 or 270	$\pi/2$ or $3\pi/2$

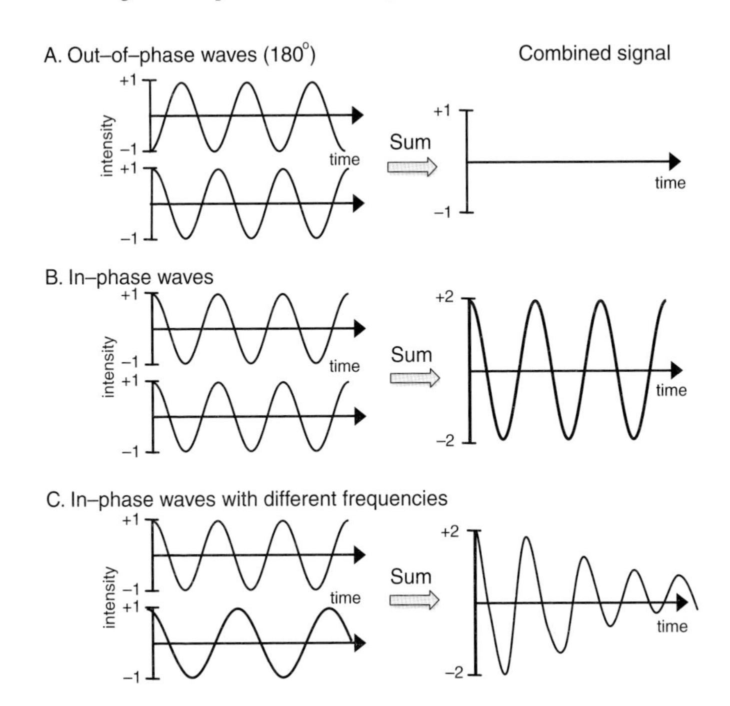

Fig. 4.16 (**a**) When two waves are 180° *out of phase*, they cancel each other out and combine to give no signal. (**b**) In contrast, when two waves are *in phase*, they combine to produce a signal with an intensity equal to the sum of the two waves. (**c**) When two waves have slightly different frequencies, they combine to produce a signal that decays over time

at $y = 0$ can have a phase of 90° (or $\pi/2$) or 270° (or $3\pi/2$) depending on whether the wave initially goes up or down. Two waves are *in phase* when they begin fluctuating at the same point, like the two in Fig. 4.16b. Otherwise the waves are *out of phase* (Fig. 4.16a).

Think back to our discussion in Chap. 1 about the speed skaters maximizing their contact time when the first racer gives the second racer a "push start" for her leg of the race (Fig. 1.9). Remember the second racer needed to skate around the inner rink at the same frequency as the first skater racing in the outer rink in order for the two teammates to connect with each other when the second skater enters the race (Fig. 1.9b). But matching frequencies is not enough; the skaters also need to be *in phase*. If the second racer is across the rink from her teammate when she enters the race, she will definitely miss the "push start" even if her skating frequency perfectly matches that of the first racer.

The same goes for waves. If two waves have identical frequencies but their phases are 180° apart (one wave starts at 0° while the other wave starts at 180°), their signals will completely cancel each other out, and the sum of their signals will be zero (Fig. 4.16a).

In contrast, when the waves start with the same phase (are *in phase*) and have identical frequencies, the waves overlap perfectly and their signals sum together, creating one wave with twice the amplitude of the individual waves (Fig. 4.16b). Physicists say that the waves are *coherent* or have *phase coherence*. You will hear these terms often in NMR, so be sure you understand them.

Coherence is a fabulous phenomenon, and it is essential for NMR. The electromagnetic wave produced by an individual nucleus is too weak to detect—it has a teeny-tiny amplitude. But when 10^{13} of these waves become coherent, all of their amplitudes unite to produce a wave that's easy to detect—our NMR signal!

Look again at the coherent waves in Fig. 4.16b. Their amplitudes continue to add up productively as long as the waves maintain identical frequencies. If their frequencies drift off a bit (Fig. 4.16c), then the waves' phases start to diverge and their signals no longer coincide perfectly. Now when we add up the waves, the total amplitude is less than the amplitude of the individual waves. The NMR signal is vanishing! NMR spectroscopists call this process *dephasing* or *lose of phase coherence*. Eventually their signals will cancel completely, and the net signal will be zero.

This is the essence of T_2 relaxation—the time it takes for the individual nuclei to fall out of "lockstep" or loose coherence (see Mathematical Sidebar 4.3 for more on T_2 relaxation). 10^{13} nuclei all begin ringing at the same frequency and phase (Fig. 4.3b, left), but as we collect the NMR data, the frequency of individual nuclei begin to shift and encompass a broad range of frequencies (Fig. 4.3, right). It's a bit like a beam of light from a flashlight—the light starts as a small pinpoint but spreads out as the light travels through space.

What effect does this "spreading out of frequencies" have on our NMR spectrum? Remember the NMR spectrum tells us the ringing frequency for each atom in the molecule. So if T_2 relaxation causes an atom type to shift its frequency randomly as we collect the NMR data, does this mean we get more than one peak in the NMR spectrum? Not quite; the spreading out of frequencies makes the peak *broader*. In fact, the quicker the waves lose coherence, the broader the peak. *In a nut-shell, the spreading out of frequencies, spreads out the NMR peak!* To understand why, let's look at a few examples.

Say calmodulin's 137-H_N amide hydrogens all ring at 9.104 ppm, and they all have an infinitely long T_2 relaxation constant. In other words, the NMR signal appears as *one* sine wave with *one* frequency that lasts forever (left, Fig. 4.17a). If we perform a Fourier Transform

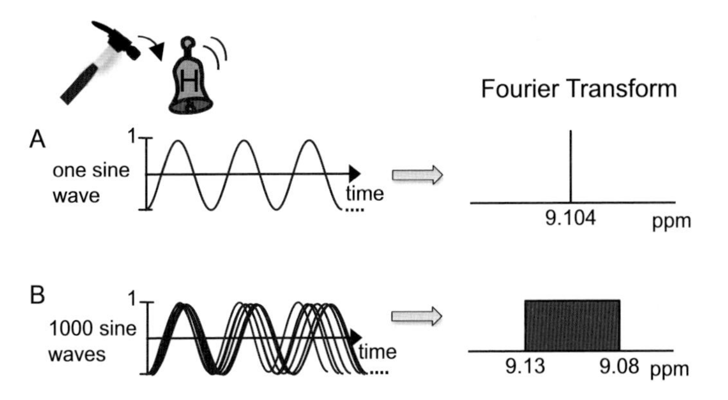

Fig. 4.17 (**a**) The Fourier Transform of an infinite sine wave is a straight line called a delta-function. (**b**) If you have a 1,000 of these sine waves all with the same amplitude and with frequencies evenly distributed between 9.08 and 9.13 ppm, then the Fourier Transform contains 1,000 of these peaks, merging together to create a plateau or square-shape peak

on this curve, what will the spectrum look like? Remember the Fourier Transform is a merely "frequency finder"—it tells us the frequency of each sine wave present in the NMR signal. This particular curve has one sine wave at 9.104 ppm, so we get one razor-thin peak right at 9.104 ppm (right, Fig. 4.17a).

Now imagine the other extreme. Say that each of 10^{13} 137-H_N hydrogens rings at one of 1,000 different frequencies between 9.08 and 9.13 ppm (left, Fig. 4.17b). The Fourier Transform will find all 1,000 frequencies and give 1,000 individual peaks. However, if the frequencies are too close to each other, their peaks will blend together and produce one broad plateau that encompasses all 1,000 frequencies. In addition, the height of the plateau at a particular frequency will tell us how well that frequency is represented in the ensemble of waves. In this example, any frequency between 9.08 and 9.13 ppm occurs with the same likelihood, so the peak is completely flat (right, Fig. 4.17b).

In reality, atomic nuclei behave as a mixture of these two extremes (Figs. 4.17a versus 4.17b): They start off like the first case with all 10^{13} 137-H_N nuclei ringing together at the same frequency, and they end up like case two with each nucleus ringing at a slightly different frequency.

The Fourier Transform is super smart; you can't trick it. It will catch all these frequencies even though they shift over time. It will find the original ringing frequency (9.104 ppm) as well as the 1,000 other frequencies appearing at the end of our NMR signal. The result is a stretched-out peak centered around 9.104 ppm called a *Lorentzian* shape (Fig. 4.18a, right).

The Lorentzian shape is quite beautiful. It has a "pointy" center at the maximum frequency (9.104 ppm for the 137-H_N atoms) and gentle "wings" that spread out symmetrically on either side (4.18a, right), representing the frequencies sampled by the waves as they dephase or lose coherence. The width of these "wings" depends on how long the atoms stay in tune: atoms with long T_2 values that ring at the same frequency for quite a long time (atoms that stay in tune), will have a

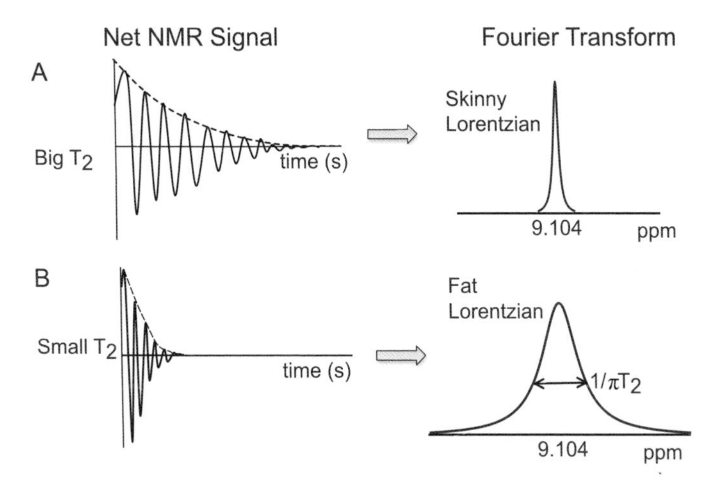

Fig. 4.18 The Fourier Transforms of an exponentially decaying sine wave (left) is a peak with a Lorentzian shape (right) (**a**) Atoms with NMR signals that slowly decay (large T_2 values) create skinny Lorentzian peaks. (**b**) In contrast, atoms with NMR signals that rapidly decay (small T_2 values) create fat or broad Lorentzian peaks

skinny Lorentzian peak (Fig. 4.18a). In contrast, atoms with short T_2 values (atoms that quickly go out of tune) will have a broad peak that is more similar to the plateau (Fig. 4.18b).

We'll learn more about these peak "shapes" and "widths" in the next few chapters, but the key point to remember is that the T_2 relaxation time tells us how broad our NMR peaks are. Short T_2 values give us fat peaks, and long T_2 values give us thin peaks. Specifically, the peak width (at half-height) is equal to $1/T_2\pi$. We need the factor of π because T_2 is in units of time (seconds) but the peak's width is in units of frequency (Hz). Don't worry about these details; just remember that fast decaying signals give fat peaks and slowly decaying signals give skinny peaks.

Mathematical Sidebar 4.3: T_2 Relaxation and Spin Echoes

T_2 measures the time it takes for nuclei to "dephase" or for individual nuclei to get "out of sync" with each other. At the start of an experiment, the nuclei are "lined up" with the magnetic field (along the *z*-axis), creating a net magnetization I_z. When we "hit" these nuclei with an electromagnetic field in the *xy*-plane, they move into the *xy*-plane and create our NMR signal. At that instant, all the nuclei are in phase (i.e., they are coherent), and our NMR signal is strong. However, due to small differences in their resonance frequencies, they will immediately begin to dephase, reducing the intensity of the signal that we can observe in the *xy*-plane. As these spins continue to spread out along the *xy*-plane, we see an oscillation in the intensity observed along the *x*-axis. The rate at which the coherence is lost is given by a first-order rate expression:

$$\frac{\mathrm{d}}{\mathrm{d}t}I = \frac{-I}{T_2^*}$$

Where I is the signal intensity at a particular time t, and $T_2{}^*$ is the apparent transverse relaxation time.

This is called the "apparent" relaxation time because in addition to small differences in the local magnetic field caused by the other nuclei and motion, external factors can also decrease coherence. In particular, slight variations in the external magnetic field along the *z*-axis (i.e., an inhomogeneous B_o) cause nuclei in the same NMR sample to "feel" a different magnetic field, leading to variations in their ringing frequency and faster dephasing. A spin-echo experiment, as described in Chap. 6 (Sect. 6.6 and Fig. 6.4), can separate the contributions to T_2 due to an inhomogeneous magnetic field from factors intrinsic to the nuclei. In this experiment the nuclei are allowed to dephase for a time t

during which the nuclei "fan out" as some spins move toward the $+x$-axis more quickly than others (Fig. 6.4a). Then the nuclei are "hit" with a second pulse, a 180° pulse. This "flips" all the spins (Fig. 6.4b) but does not change their speeds. The spins that were "travelling" at a faster rate are now farther from the $-x$-axis than the slower spins. After another time t, the spins should all "rephase" and be together along the $-x$-axis (Fig. 6.4d). Thus, this 180° pulse "refocuses" nuclei that "feel" different magnetic fields and removes the effects of B_0 inhomegeneity. The intrinsic transverse relaxation T_2 is measured by taking the difference in the signal intensity along $-x$-axis right after the first pulse moves the spins to the xy-plane (Fig. 6.4a, left) from the signal intensity along the $-x$-axis after the nuclei are refocused (Fig. 6.4d, left).

References and Further Reading

Cantor CR, Schimmel PR (1980) Biophysical chemistry part II: techniques for the study of biological structure and function, Chap. 9. W. H. Freeman and Company, New York.

Cavanagh J, Fairbrother WJ, Palmer AG III, Rance M, Skeleton NJ (2007) Protein NMR spectroscopy: principles and practice, 2nd edn., Chaps. 1 and 5. Academic Press, Amerstdam.

Grzesiek S (2007) Notes on relaxation and dynamics from EMBO Practical Course on NMR, Basel, September 7–14.

Levitt M (2001) Spin dynamics: basics of nuclear magnetic resonance, Chaps. 7 and 16. John Wiley & Sons, Inc., Chichester.

Chapter 5
Relaxation Theory Part Two: Moving Atoms and Changing Notes

Let's try a simple thought experiment. Imagine staring at a red light two feet in front of your face. The light stays red for a few seconds and then instantly changes to green (Fig. 5.1a). A few seconds later, it's back to red. What do you see? Red and green flashing lights, of course—nothing interesting about that.

Okay, now imagine that the light switches from red to green extremely fast, like once every few milliseconds. What would you see?

It takes the light about 100 ms to travel up your optic nerve to the cerebral cortex where the photons get processed into electrical signals that represent the color red in your thoughts. Because the lights are now flashing between green and red faster than your brain can process, the signal gets blended and summed up. Instead of seeing two flashing lights, you now see only one light—a yellow light (Fig. 5.1b)! (Remember that red and green lights combine together to create yellow.)

Now think of the other extreme: the two lights take a very long time to change colors, like 20 minutes. Unless you sit there and stare at the lights, all you see now is one color—red (Fig. 5.1c).

Indeed, the human eye is tuned to see the blinking only over a certain range of frequencies—flashings that occur over seconds, maybe even minutes. But when the lights blink significantly faster, the two colors merge into one (Fig. 5.1b). And, when the lights flash much slower, we just don't see the blinking at all.

M. Doucleff et al., *Pocket Guide to Biomolecular NMR*,

DOI 10.1007/978-3-642-16251-0_5,

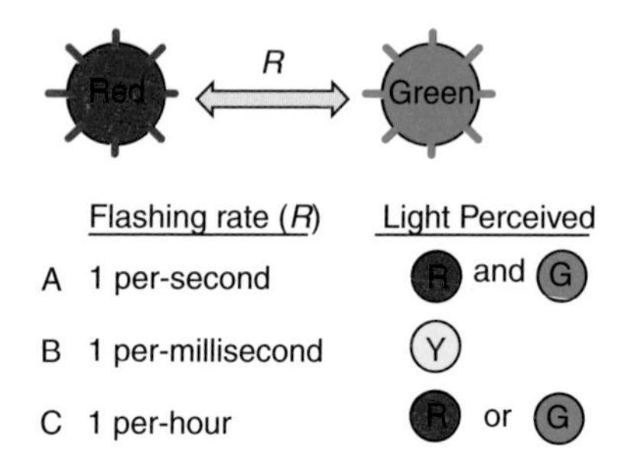

Fig. 5.1 When a light is flashing between red and green, the color perceived by your eyes depends on the rate at which the light flashes. (**a**) If the light flashes one time each second, then you see both colors, red and green. (**b**) If the light flashes very quickly, like one time each millisecond, then you see only one color light, yellow. (**c**) If the light flashes very slowly, like one time each hour, then you most likely only see one color too, either red or green

NMR is quite similar—it can "see" or detect motions that occur only over a range of particular rates or frequencies. When the motion between two states occurs significantly more often, then states get averaged into one state; when the motion occurs significantly slower, NMR can't detect it (and simply detects two states)! Luckily for us, NMR can "see" a large range of motional rates, and this range overlaps nicely with the rates of motion that are important for protein dynamics and catalysis. In the following chapter, we'll explore these "time scales" and see how we can use NMR to characterize a vast array of biological motions.

5.1 Keeping the Terms Straight

To understand protein dynamics analysis by NMR, you need to know three terms:

(a) Time scales
(b) Exchange Rate
(c) Motional Averaging

We'll tackle the first two in this section, and cover the third one in the following sections.

The motion of proteins and other macromolecules can be quite complicated: dozens of atoms and bonds rearranging to accommodate substrates; loops opening and closing around active sites; even entire protein domains can fluctuate around hinge points or flexible linkers to make large conformational changes. In most cases, luckily, we can break the complex dynamics into an extremely simple model, where one atom in the protein moves back-and-forth between two locations (say A to point B) at a particular frequency (Fig. 5.2, top). NMR spectroscopists call this back-and-forth motion *exchange*. How often the exchange occurs is called the *exchange rate* and the inverse of the rate (1/rate) is called the *time scale* of the motion. For example, if the atom moves back-and-forth between A and B 1,000 times per second, then

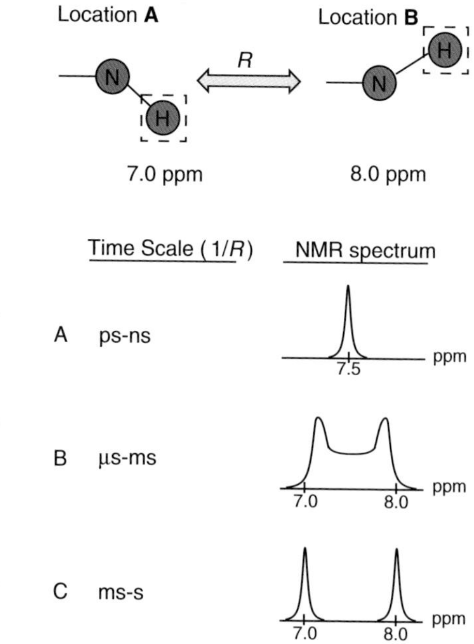

Fig. 5.2 The hydrogen atom moves between two locations, A and B, where the chemical shift at each location is 7.0 and 8.0 ppm, respectively. When the rate of exchange, *R*, is (**a**) in the fast time regime, we see one peak in the NMR spectrum; (**b**) in the intermediate time regime, we see one distorted peak in the NMR spectrum; and (**c**) in the slow time regime, we see two peaks in the NMR spectrum

the exchange rate is 1,000 s^{-1}, and the time scale of the motion is 1/1,000 of a second (i.e., 1 millisecond).

Biomolecular NMR spectroscopists and protein biochemists love the word "time scale," And, they usually use it when they want to specify a range of times in which a particular event happens, such as "picosecond to nanosecond time scale" or "millisecond to microsecond time scale."

For example, think of a firefly bating its wings: it flaps approximately 10 times per second. But the precise rate depends on the species of firefly, the temperature outside, and how much sugar the firefly has consumed. However, no matter the conditions, the exchange rate is somewhere between 10 and 100 times per second. Thus, NMR spectroscopists would say the wing flapping "occurs on the millisecond to second time scale" (a time scale too fast to be detected by our eyes, so the wings appear blurry or smeared out!).

In proteins, hydrogen atoms like to vibrate about their bonds, and this happens extremely often (Fig. 5.3a), like once every 10 picoseconds, or on "the picosecond to nanosecond time scale." On the other hand, the motion of loops and domains of proteins, usually occurs less frequently, like once every 100 microseconds, so loop motion often occurs on the "microsecond to millisecond time scale" (Fig. 5.3b). Intermolecular interactions in biology are often even slower than that. For example, many DNA-binding proteins hop on and off DNA-binding sites only few times per second (Fig. 5.3c). What is the time scale for this motion? Millisecond to second time scale, of course!

You probably get the idea. The key point is that the exchange rate tells how frequently the motion occurs, *not* how fast the molecule moves when it travels between points A and B. The point is subtle but extremely important. When the hydrogen goes from point A to point B (Fig. 5.2), it does so quite rapidly, essentially instantaneously. The time scale simply tells us how often the hydrogen decides to make the trip. Think back to the flashing lights (Fig. 5.1). In all three cases, the

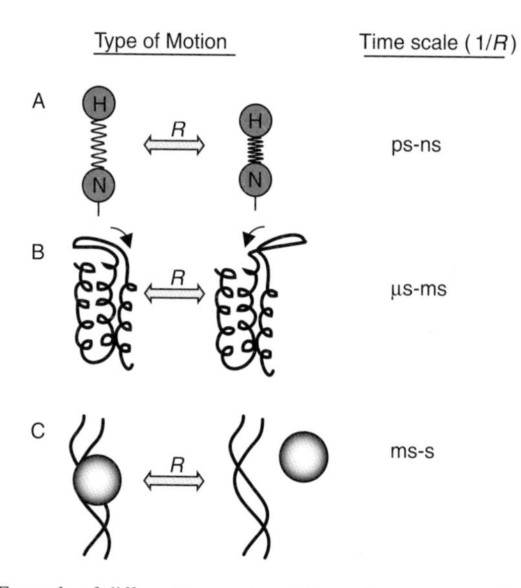

Fig. 5.3 Example of different types of protein motion on various time scales. (**a**) Atoms vibrate about a bond on the picosecond-to-nanosecond time scale; (**b**) loops on proteins often change configuration on the microsecond-to-millisecond time scale; and (**c**) proteins (gray circle) often hop on-and-off DNA (black lines) on the millisecond-to-second time scale

light changes color instantaneously and what actually determined the color detected by your eyes is how often or how frequently the light blinks.

5.2 NMR Dynamics in a Nutshell: The Rules of Exchange

In the next few chapters of the book we will cover the details for how NMR helps to characterize molecular motions that occur on many different time scales. But before we dive into the deep-end of this NMR pool, let's wade in the shallow-end for awhile and outline a few general principles for the simple model of a hydrogen atom going from location A to location B at a particular exchange rate (Fig. 5.2).

Say that the hydrogen has a chemical shift of 7.0 ppm when it's in location A but a chemical shift of 8.0 ppm at location B. What do we see in the NMR spectrum for this hydrogen? Well, it all depends on how often the hydrogen makes the trip from A to B:

(a) If the exchange rate between the locations is *extremely fast* (such as picosecond to nanosecond time scale) (Fig. 5.2a), you will see ONE peak for the hydrogen somewhere between 7.0 and 8.0 ppm. Spectroscopists call this case "fast exchange." We will explore this more in the next section.
(b) If the exchange between the two locations is extremely slow (such as the millisecond to second time scale) (Fig. 5.2c), you will see TWO peaks for the hydrogen, one at 7.0 ppm and one at 8.0 ppm. NMR spectroscopists call this "slow exchange." We will talk about it more in Sect. 5.4.
(c) If the exchange between the locations is somewhere in the middle (like microsecond to millisecond time scale) (Fig. 5.2b), the situation gets more complicated, but in general, you will see one stretched-out ("broadened") peak that may even have a funny shape, like two peaks smeared together. NMR spectroscopists call this "intermediate exchange." We will talk more about this in Sect. 5.5.

These are the most important three "exchange regimes" in NMR dynamics studies, and they will take you surprisingly far when you're reading the NMR literature. To simplify further, just remember:

> Frequent motion gives one peak; infrequent motion gives two peaks; and when the motion occurs at an intermediate frequency, you get a peak that's distorted

Now let's see why this is true!

5.3 Two States, One Peak: Atoms in the Fast Lane of Exchange

All this talk of frequencies must remind you of something—no, not the REM song, "What's the Frequency, Kenneth!"—rather, the ringing frequency of a nucleus (Fig. 5.4). Indeed, just as motion has a time scale, the ringing frequency also has its own time scale, approximately the reciprocal of the ringing frequency (in Hz). NMR spectroscopists call this the *Larmor time scale* (because the ringing frequency in radians-per-second is called the Larmor frequency). So for hydrogens in a 11.75-Tesla magnet that ring at approximately 500 MHz (or 500×10^6 cycles-per-second), the "ringing" time scale is $1/500 \times 10^6$ Hz seconds ~2 ns or the nanosecond time scale.

This means that it takes approximate 2 ns for the nucleus's little magnet to rotate around the z-axis one time (Fig. 4.7; Sect. 4.3). For the Fourier Transform to find the frequency at which the hydrogen rotates, we need the little magnet to make at least a few cycles (otherwise the Fourier Transform won't have enough information to identify the precise frequency). So what happens if the hydrogen exchanges between location A and B many times before the nuclei makes one cycle around the z-axis?

Think about it. If the frequency shifts back-for-forth from 7.0 to 8.0 ppm many times within 2 ns, then the spectrometer won't have time to measure two frequencies separately. Instead, the hydrogen will appear to be ringing at a single frequency somewhere between 7.0 and 8.0 ppm.

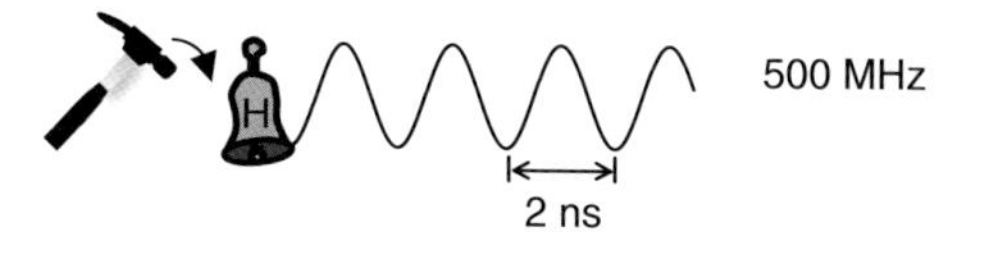

Fig. 5.4 For a hydrogen nucleus ringing at 500 MHz, it takes 2 ns for the nucleus to make one rotation around the z-axis. This precession period defines the *Larmor time scale*

It is quite similar to your brain blending the red and green lights into yellow when it doesn't have time to process the two signals separately. The spectrometer can't keep up with the shifting frequencies, just as your eyes can't keep up with the blinking lights! The result is one peak in the NMR spectrum somewhere between 7.0 and 8.0 ppm (Fig. 5.2a). Again, this is known as the "fast exchange" regime.

The exact chemical shift of the peak depends on the time that the hydrogen stays each frequency. For example, say that the hydrogen spends 90% of the time (on average) at location B and the remaining 10% of the time in location A (Fig. 5.5a). What does the NMR spectrum look like?

To figure this out, let's pretend that the hydrogen is a jogger running around a track (Fig. 5.6), and the Fourier Transform is its coach recording the hydrogen's pace every time it passes the starting line. If the runner spends 90% of the time running at an 8 minutes-per-mile pace and 10% of the time at a 7 minutes-per-mile pace, what pace will

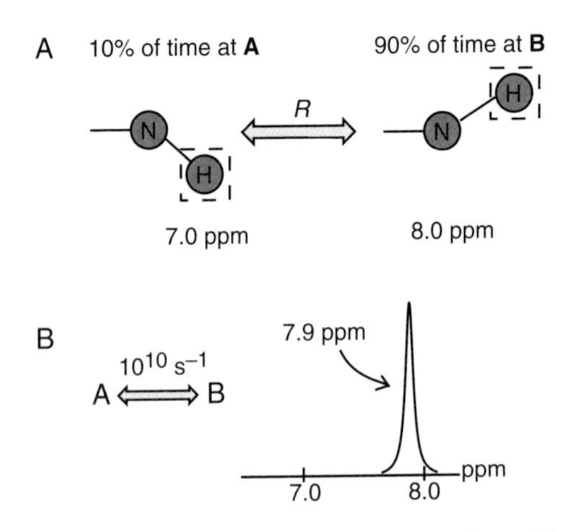

Fig. 5.5 (**a**) If a hydrogen atom spends 10% of the time ringing at 7.0 ppm (location A), 90% of time ringing at 8.0 ppm (location B), and exchanges between the two locations extremely rapidly (10^{10} s^{-1}), (**b**) then the NMR spectrum will contain one peak with a chemical shift of 7.9 ppm

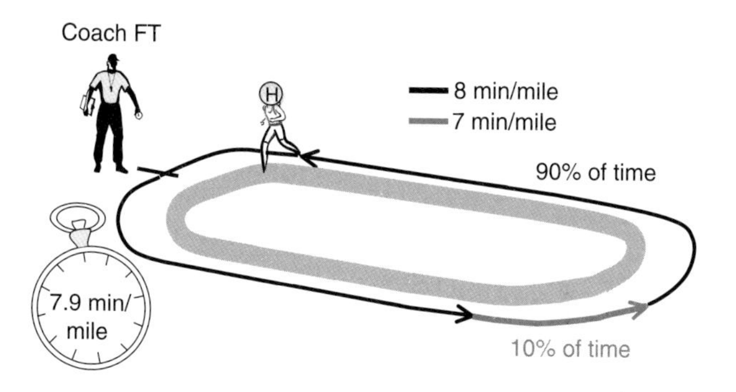

Fig. 5.6 If a runner's pace is 7 minutes-per-mile for 10% of the time, 8 minutes-per-mile for 90% of the time, and the runner changes pace many times during each lap, then the coach records a 7.9 minutes-per-mile pace for *every* lap

coach Fourier record for each lap? Remember that we are in the "fast exchange regime," which means that the hydrogen will shift its pace many times during each lap, so the average pace recorded by coach Fourier will be the same for *every* lap. And, each time it will simply be the "weighted average" of the two speeds:

$$(7 \text{ minutes-per-mile}) \times 0.1 + (8 \text{ minutes-per-mile}) \times 0.9 = 7.9 \text{ minutes-per-mile}$$

The hydrogen shifting between ringing frequencies of 7 and 8 ppm in the NMR experiment is exactly the same: When the hydrogen shifts between the ringing frequencies many times during one precession around the z-axis, the peak in the NMR spectrum will be the weighted average of the two chemical shifts, depending on how long it spends at each location (Fig. 5.5b):

$$(7.0\,\text{ppm}) \times 0.1 + (8.0\,\text{ppm}) \times 0.9 = 7.9\,\text{ppm}.$$

NMR spectroscopists call this phenomenon "motional averaging" because the dynamics average the NMR signals for the two states of the hydrogen into one resonance frequency (and thus one peak in the spectrum).

That makes reasonable sense, but what about "loss of phase coherence?" In the previous chapter, we learned that a slight shift in frequencies causes the signal to decay because the millions of identical hydrogens fall out of synchrony as they spin around the *z*-axis. Does that happen here? Surprisingly no. Because the nucleus jumps back and forth between 7.0 and 8.0 ppm so quickly, phase differences don't have time to accumulate in either location. So our signal stays nice and strong despite the changing frequencies.

One last note before we move on: Instead of referring to the ringing frequency and ringing time scale, NMR spectroscopists will often use what's known as the *Larmor frequency or Larmor time scale*. The *Larmor frequency* is simply the ringing frequency in units of radians-per-second instead of cycles-per-second (Hz). Since there are 2π radians in each cycle, we simply convert the ringing frequency to Larmor frequency by multiplying the Hz by 2π, and to convert the ringing time scale to the Larmor time scale, we simply divide by 2π. You don't need to worry too much about these details; just remember that the Larmor (or ringing) time scale is on the picosecond to nanosecond time scale for hydrogens.

5.4 Two States, Two Peaks: Atoms in the Slow Lane of Exchange

Now let's talk about the other extreme—when the hydrogen exchanges between locations A and B super slowly, say like once every 10–20 seconds (or the seconds to minutes time scale). What does the NMR spectrum look like now?

In Chap. 4, we learned that the NMR signal doesn't last forever but instead decays exponentially and vanishes after approximately 0.5 seconds (Fig. 5.7). This means we have less than a second to detect the hydrogen jumping between locations A and B. When the exchange rarely occurs during this detection period, the hydrogen will "look" like

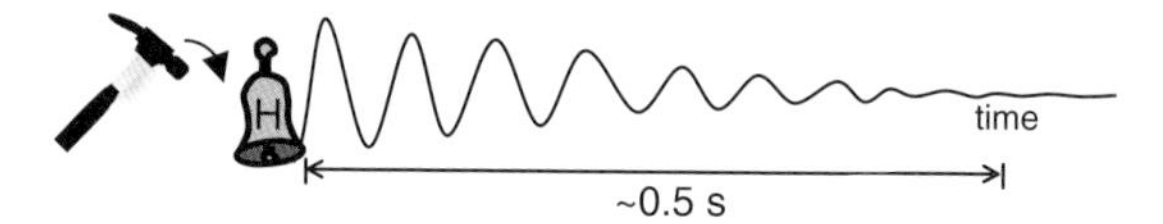

Fig. 5.7 The NMR signal for hydrogen atoms in proteins usually vanishes after 0.5–1 second. This decay time defines the *relaxation time scale*

its standing still. If the hydrogen is in location A, the hydrogen will ring at 7.0 ppm; if it happens to be at location B, it will ring at 8.0 ppm. Since we are detecting billions of these specific types of hydrogens, the NMR spectrum will then have two peaks, one at 7.0 ppm and one at 8.0 ppm (Fig. 5.2c).

In the previous chapter, we learned that the R_2 or *transverse relaxation rate* determines how quickly the NMR signal decays. Like all rates and frequencies, this relaxation rate also has a time scale—approximately $1/R_2$ (~1/0.05 seconds) or the seconds time scale. So when the exchange between locations A and B is significantly slower than seconds, NMR spectroscopists say that the "exchange is slow compared to the relaxation time scale." In this case, we see two peaks in the spectrum (Fig. 5.2c).

If that all doesn't make sense, think back to the running analogy. Now the hydrogen rarely changes its pace even after multiple laps around the track. So coach Fourier Transform will record either a 7 minutes-per-mile pace or an 8 minutes-per-mile pace and he will rarely record any pace in between the two. The result is two distinct paces, one at 7 minutes-per-mile and one at 8 minutes-per-mile.

5.5 Two States, One Strange Peak: Atoms in Intermediate Exchange

Okay, so when the hydrogen moves between locations A and B with a rate that's slow compared to the decay time of NMR signal (i.e., "the exchange is slow on the relaxation time scale"), then we see two peaks in the NMR spectrum. But when the rate is faster than the ringing frequency (i.e., "the exchange is fast on the Larmor time scale"), we see one peak in the NMR spectrum. That all makes sense, right? But what about when the exchange rate falls somewhere between the cases—slower than the Larmor time scale but faster than the relaxation time scale (Fig. 5.2b)? Before we answer that question, we need to learn about one more time scale: the *chemical shift time scale.*

In Chap. 4, we learned that when nuclei shift their ringing frequencies, their NMR peak becomes broader because the Fourier Transform identifies multiple frequencies in the NMR signal (Sect. 4.5 and Fig. 4.18). Furthermore, the larger the shift in ringing frequency, the broader its peak. The same idea holds true for an atom exchanging between two locations with different ringing frequencies. When an atom moves location while we collect its NMR signal (Fig. 5.2, top), its ringing frequency will also shift. Most importantly, the amount that the frequency changes depends on the difference in the ringing frequency at the two locations. When this difference is very small, say only 1 Hz, then moving from locations A and B will hardly alter the ringing frequency (or the phase of the signal) (Fig. 5.8). The nuclei's coherence is preserved, and we still observe a beautiful sharp peak in the spectrum (because the Fourier Transform can still identify a distinct sine wave in the signal). However, when the difference in ringing frequency at the locations A

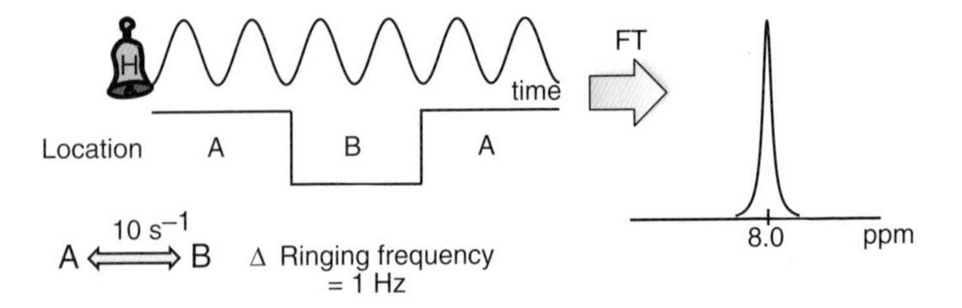

Fig. 5.8 If the difference in ringing frequency at locations A and B is very small, like 1 Hz, then we find only one peak in the NMR spectrum, even if the hydrogen atom exchanges between the two locations very slowly, like 10 times-per-second

and B is quite large (e.g., 1,000 Hz), then the atom's frequency will shift quite quickly when it moves locations. The nuclei will quickly loose coherence, and the peak in the NMR spectrum will broaden out.

Clearly, *the difference in the ringing frequency at the two locations* is the most important factor for determining what we see in the NMR spectrum when an atom moves between two locations at a particular exchange rate. This is the essence of the chemical shift time scale. Specifically, the inverse of the frequency difference between states (i.e., 1/change in ringing frequency) is called the *chemical shift time scale.* This is probably the most important time scale in NMR, so be sure you understand it well!

So what is the chemical shift time scale for the hydrogen in Fig. 5.2, where the chemical shift difference between locations A and B is 1 ppm? Well, it depends on the strength of the external magnetic field. At 11.75 Tesla (or 500 MHz), the change in ringing frequency between locations A and B is 500 Hz (because 1 ppm = 500,000,000 Hz/1,000,000 = 500 Hz). So the chemical shift time scale is approximately $1/2\pi(500) = 0.3$ ms, or the microsecond to millisecond time scale. Again, the factor of 2π comes into play because we need to convert the frequency in Hz (cycles-per-second) to radians-per-second before taking the reciprocal.

The chemical shift time scale is what actually divides the motion into the three categories shown in Fig. 5.2. Specifically:

1. An exchange rate that is more than about *10 times* the difference in the ringing frequencies between the two locations is in *fast exchange on the chemical shift time scale.*
2. An exchange rate that is less than about *10 times* the difference in the ringing frequencies between the two locations is in *slow exchange on the chemical shift time scale.*
3. An exchange rate that lies somewhere between the first two cases is in *intermediate exchange on the chemical shift time scale.*

We know what the NMR spectrum looks like for cases 1 and 2 (that's easy, right? one peak for case 1 (Fig. 5.2a) and two peaks for case 2

(Fig. 5.2c). But what about for case 3? It's actually intuitive—you see a combination of cases 1 and 2. In other words, you see one peak spread out between the ringing frequencies at the two locations (Fig. 5.2b). Let's see why.

For case 3 above, the rate of exchange between the two locations is comparable to the difference in the ringing frequencies between locations A and B. Now the hydrogen switches between the two locations many times during our detection of the NMR signal (Fig. 5.9a). And, the frequencies at the two locations are different enough that the phase of the signal (i.e., its position or angle in the sine curve; Sect. 4.5 and Table 4.2) changes significantly when the hydrogen repositions itself (Fig. 5.9a). This will definitely cause our NMR signal to decay more quickly (Fig. 5.9b), but what on Earth does the spectrum look like now?

Think again of the running analogy, except now let's consider the pace in terms of minutes-per-lap instead of minutes-per-mile to make the analogy clearer. Say that the hydrogen starts off with a 1 minute-per-lap pace (or 1 lap-per-minute frequency) but there is no way that it can maintain this world-record speed. So it switches to a more reasonable 2 minutes-per-lap pace every 1.5 minutes *on average* (sometimes it switches more often and sometimes less often, but on average, every 1.5 minutes). Now the hydrogen's running falls into category 3—its "intermediate exchange" on the lap-running time scale. What does coach Fourier record?

The first time around the track, the hydrogen switches to the slower pace right before it finishes the loop, so coach Fourier will record a pace slightly slower than 1 minute-per-lap, like 1.1 minutes-per-lap (Fig. 5.10a). On its second lap, the hydrogen almost makes it around an entire lap at the slower speed before changing to the faster pace, so coach Fourier will record a pace that's slow, like 1.9 minutes-per-lap. Now on his third lap, the hydrogen spends about equal time at the two paces, leading to a pace that is exactly half way between 1 and 2 minutes, or 1.5 minutes-per-mile.

You probably get the idea. Every lap has a different pace (Fig. 5.10b). If the hydrogen makes 100 laps, coach Fourier will have

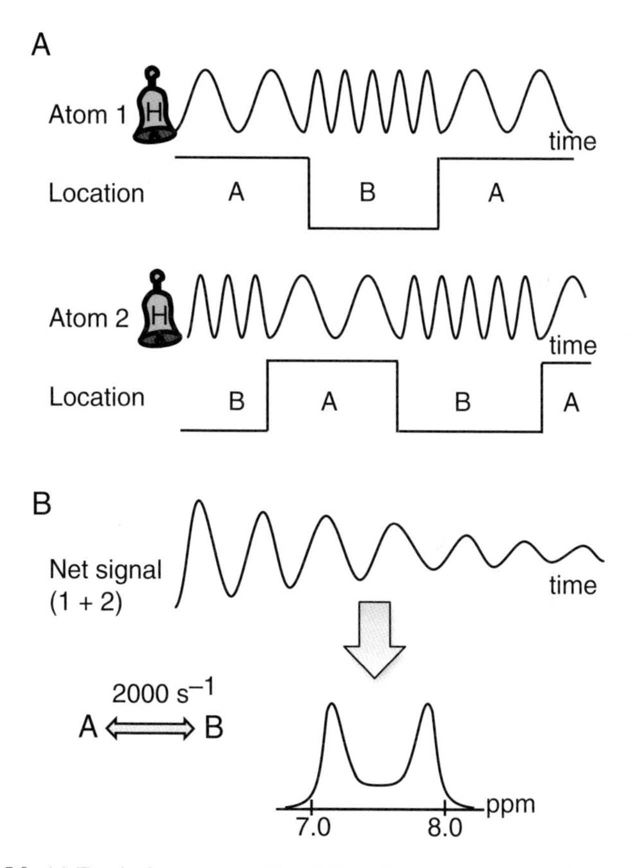

Fig. 5.9 (**a**) Two hydrogen atoms (1 and 2) exchange between locations A and B during the NMR detection period. The ringing frequency between the two locations (500 Hz) is comparable to the rate of exchange between the two locations (2,000 s^{-1}). (**b**) This causes rapid signal decay and a distorted, broad peak in the NMR spectrum

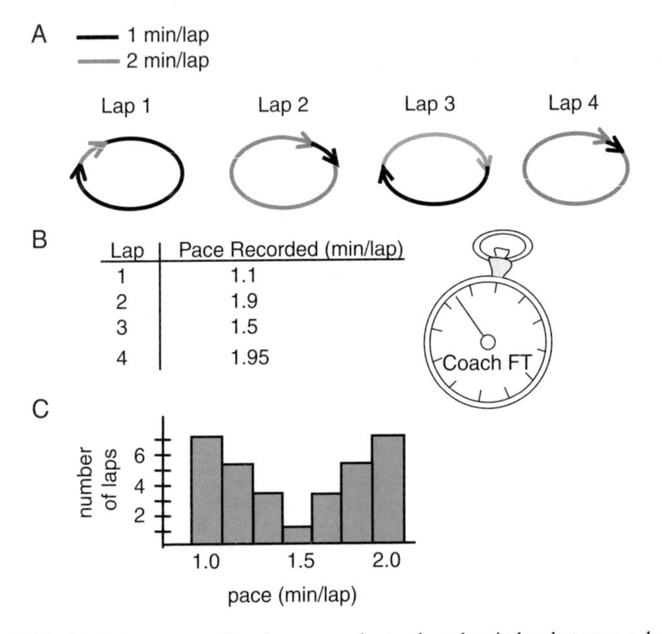

Fig. 5.10 (**a**) A jogger runs four laps around a track and switches between a 1 and 2 minutes-per-lap pace every 1.5 minutes on average. (**b**) The pace recorded by Coach Fourier is different for each lap but is always somewhere between 1 and 2 minutes-per-lap. (**c**) Thus, a histogram showing the number of laps recorded at each pace contains two broadened peaks between 1 and 2 minutes-per-lap

a list of almost 100 different paces. Many will be close to either 1 or 2 minutes-per-lap pace, but there will be many laps that fall between these extremes. If we plot a histogram of the hydrogen's paces, we see two broad peaks near 1 and 2 minutes-per-lap with significant contributions from the intermediate paces as well, like 1.5 minutes-per-mile (Fig. 5.10c).

This is quite similar to what we see in the NMR spectra when the exchange rate between the two locations A and B is comparable to the difference in the ringing frequencies at the two locations (Fig. 5.9b). In this case, the ringing difference is 500 Hz at 11.75 Teslas. If the exchange rate is approximately 2,000 s^{-1} or "on the microsecond to millisecond time scale" (because $1/2\pi(500) = 0.3$ ms), we get two peaks that are both broad and stretched-out but centered around 7.0 and 8.0 ppm (Fig. 5.2b), just as the histogram for the lap paces recorded by coach Fourier contains two stretched-out peaks near 1 and 2 minutes-per-lap (Fig. 5.10c). When this occurs, NMR spectroscopists say that the exchange rate is *intermediate on the chemical shift time scale*, or the motion is in *intermediate exchange*.

If the exchange rate speeds up significantly, then the motion begins to become *averaged-out* as discussed in Sect. 5.3, and the two broad peaks merge into one (Fig. 5.2a). In contrast, if the exchange rate slows down, the hydrogen's trips around the z-axis will start to be either at 7 or 8 ppm and rarely between the two frequencies, so the two peaks will split apart further and start to narrow (Fig. 5.2c).

You can also think of the intermediate exchange in terms of loosing coherence, as described in the previous chapter (Sect. 4.5). When the difference between the ringing frequencies is big enough and the exchange rate is fast enough, the phase of the hydrogen's atoms shift significantly with each trip around the z-axis. The specific amount of phase change varies significantly from hydrogen to hydrogen just as the running pace varied for each trip around the lap. This causes the hydrogens to "fall out of synchrony"; their NMR signals begin to cancel each other out; and the net signal detected by the spectrometer decreases or "decays" more rapidly (Fig. 5.9b). With a shorter signal and jumbled-up frequencies, the Fourier Transform can't determine the exact ringing frequency for that hydrogen and therefore returns a broad range of chemical shifts from 7.0 to 8.0 ppm (Fig. 5.9b, bottom).

5.6 Tumbling Together: Rotational Correlation Time (τ_c)

So now we understand how jumping between two locations affects our NMR spectrum. Although this type of motion is quite common in biomolecular NMR, it is not the primary way atoms in proteins and nucleic acids move around in solution. Instead of hopping nice-and-neatly between two well-defined locations, proteins, and DNA molecules like to tumble around in solution like an astronaut in space.

Think back to the pasta analogy that we used in Chap. 3 (Sect. 3.4). As the water heats up, the spaghetti adsorbs the thermal energy and begins to wiggle around in all directions. As the water gets hotter, the wiggles get faster and faster. Proteins are quite similar: The thermal energy at room temperature causes the protein to turn and rotate around in solution in all directions with no care for gravity or other preferred direction. The result is a random tumbling that is quite chaotic, complicated, and difficult to characterize.

Fortunately though, when proteins are well behaved—when they are folded into a single shape and configuration—all the atoms move together. Thus, instead of being like a floppy strand of spaghetti or fettuccini in the hot water, proteins are more like pasta shells (or *conchiglie* as the Italians call it) with all the amino acids in the protein moving as a single unit in the boiling water. When one end of the protein tumbles to the right, the other end of the protein chain always comes along with it.

This tumbling is extremely important for solution NMR because it causes fluctuating magnetic fields near atoms! Think of a hydrogen atom in your NMR sample with its little magnetic compass pointing along the huge magnetic field in the $+z$ direction, such as hydrogen B in Fig. 5.11. Remember that the hydrogen atom is not alone. It is surrounded by dozens of other hydrogen atoms, each with their own tiny magnetic field nearby (hydrogen A in Fig. 5.11). When the protein begins tumbling around in solution, the direction and the strength of these nearby fields changes continuously, causing the ringing frequency to shift constantly, as well. For example, in Fig. 5.11, hydrogen A's magnetic field is initially above hydrogen B. But after the molecule rotates, the field from A is then on the left of hydrogen B. This changes the local magnetic field felt by hydrogen B and, therefore, causes its ringing frequency to fluctuate as the molecule tumbles.

How does this frequency shifting affect the NMR signal and spectrum? As we just learned in the previous sections, the effect depends

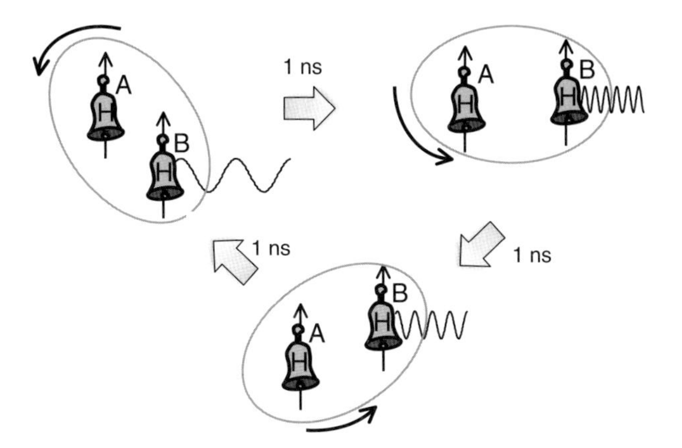

Fig. 5.11 As the molecule tumbles, the ringing frequency for each hydrogen atom shifts around due to the changes in the strength and the direction of the local magnetic field created by nearby nuclei and electrons

on the how often this frequency jumping occurs and how this rate compares to the difference in ringing frequencies sampled by the atom during the tumbling.

If the tumbling is relatively fast, the motion will average out all the ringing frequencies (Sect. 5.3 above), and we will get one sharp, beautiful peak in the NMR spectrum that represents the average chemical shift for all orientations that the molecules samples while tumbling (Fig. 5.2a). But when the tumbling rate slows down considerably and creeps into the intermediate exchange rate regime (Sect. 5.5 above), then the fluctuating fields begin to have a detrimental impact on the NMR signal. The irregular ringing frequencies cause the NMR signal for individual hydrogen atoms to fall out of synch with each other, causing T_2 dephasing to increase. The total NMR signal decays more quickly, and the Fourier Transform returns a broad peak (Fig. 5.9b, bottom) that represents all the different frequencies caused by tumbling. NMR spectrscopists call this *motional broadening*.

Unfortunately, motional broadening is usually what occurs for large molecules, like proteins and nucleic acids (>2 kDa)—their tumbling rate is slow enough that it increases T_2 relaxation and significantly broadens the peak in the NMR experiment. In fact, for proteins with any girth at all, say >100 kDa, the tumbling slows down so much that the peaks broaden beyond detection! This is why NMR is limited to only small, perky proteins that tumble quickly in solution.

So what is the time scale of this tumbling motion? Well, it depends on a few factors:

(a) Size matters most! A big, fat pasta shell stuffed with cheese or potato will definitely rotate around much slower in a pot of hot water than tiny shell pasta. The same goes for the molecules in solution: small, petite peptides tumble faster than large, bulky proteins and nucleic acids.
(b) Another obvious factor is the viscosity of the solution. If you cook pasta in a thick creamy broth, those shells will definitely tumble around much slower than when they're cooking regular in water. The same goes for the protein; put your protein in a 50% solution of thick, gooey glycerol, and it will move much slower than in a simple salt buffer. (What effect does this have on your peaks in your NMR spectrum?)
(c) Last, but definitely not least, the temperature matters substantially. Heating water dramatically decreases its viscosity, making pasta shells and proteins tumble faster at higher temperatures. (Can you figure out what effect this has on the peaks in your NMR spectrum?)

Under normal conditions—a simple salt buffer at room temperature—you can approximate the rotational time scale in nanoseconds for a protein by taking the molecular weight of the molecule in kDa and dividing by two. For example, the calmodulin protein is 17 kDa, so its rotational time scale is approximately 8.5 ns. NMR spectroscopists call this value τ_c, and you can think of it as *the time it takes for the entire protein to rotate approximately 60° in the solution.*

For NMR spectroscopists, τ_c has a precise mathematical meaning, called the *overall rotational correlation time* or *overall rotational correlation constant*. Although the definition of the rotational correlation time is extremely complicated (we'll learn more about it in the next chapter), we can get a good idea of its meaning with a simplified version.

Imagine sitting on top of a protein in an NMR sample (Fig. 5.12a). Say that your position starts along the z-axis at +1 and then the molecules begins to tumble in solution. If we plot your position along the z-axis in terms of time, we would see a chaotic curve that oscillates up and down as you roll around in solution with the molecule (Fig. 5.12a, right): as you fall down to $z = -1$ and then back up to

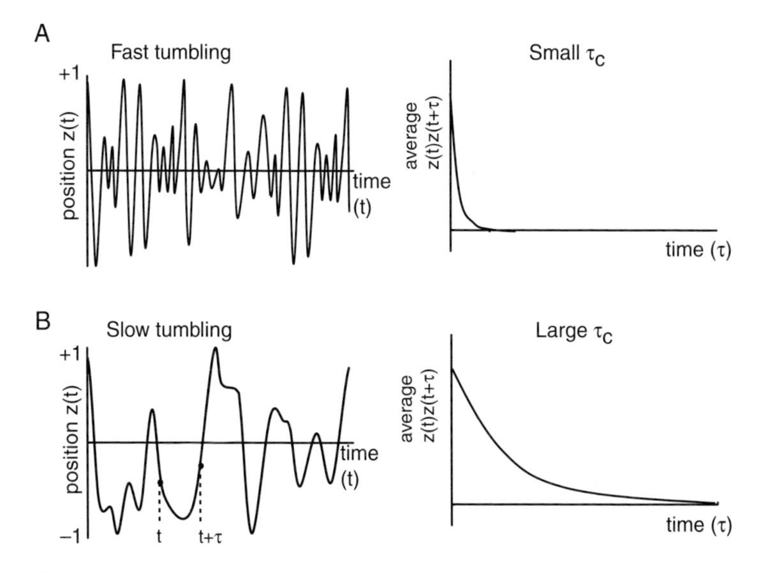

Fig. 5.12 The position of a hydrogen in the z-dimension ($z(t)$) fluctuates up and down over time as the molecule tumbles in solution. (**a**) When the molecule rotates quickly, the average position over time (and $z(t)z(t + \tau)$) quickly goes to zero and τ_c is small. (**b**) In contrast, when the molecule rotates quickly, the average position over time (and $z(t)z(t+\tau)$) decays slowly and τ_c is large. See mathematical sidebar in Chap. 6 to learn about the mathematics behind the correlation constant, τ_C, and $z(t)\, z\,(t + \tau)$

$z = +1$, you often sample every position in between. The key point is that the motion is essentially random. So over time your *average* position along the z-axis becomes zero because all the ups and downs cancel each other out.

The rotational correlation time simply tells us how long it takes this "averaging to zero" to occur. If the protein is tumbling super fast (because it is very small or in extremely hot water), then your position along the z-axis oscillates rapidly, and the average position quickly becomes zero. So the rotational correlation time is small, like a fraction of a nanosecond (Fig. 5.12a, right). In contrast, if the protein is big and bulky, it tumbles leisurely (Fig. 5.12b, left). Your average position along the z-axis changes slowly, and it takes a much longer time for it to average to zero; thus, the rotational correlation time is longer, like 20 ns (Fig. 5.12b, right).

"Correlation times" are very common in NMR spectroscopy and physical chemistry in general, so we'll learn a bit more about them in the next chapter. However, all you need to remember is that the correlation time (or correlation constant), τ, simply tells you how fast a variable repeats itself. If the oscillations are fast, the correlation time is small (Fig. 5.12a); if the oscillations are slow, then the correlation time is big (Fig. 5.12b).

To help remember this, think back to sitting on top of a protein. If you are moving fast, your correlation time is small because you quickly return to your starting position. In contrast, if you're moving slowly, your correlation time is large because it takes a long time to return to your starting position. To understand more of the mathematics behind correlation constants, check out the mathematical sidebar in Chap. 6.

5.7 Summary

This chapter is jam-packed with new terms and definitions that are used constantly throughout journal articles on NMR dynamics. So to help you keep track of it all, Fig. 5.13 provides a summary of the various time scales present in biomolecular dynamics and their effect on NMR spectra.

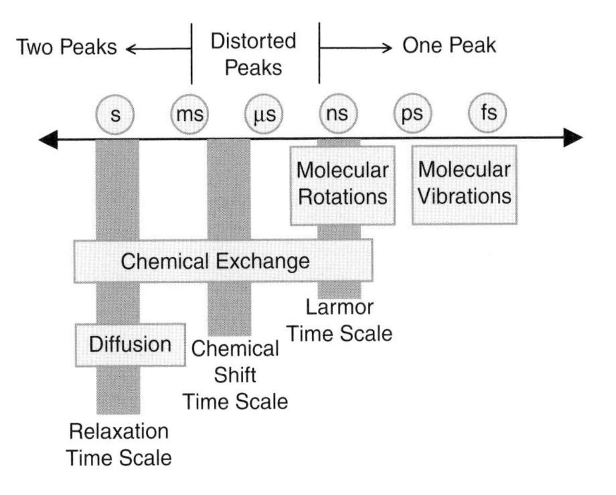

Fig. 5.13 A summary of the time scales for various types of molecular motion in proteins and the NMR time scales with which they overlap. The top of the diagram describes the effect these dynamics usually have on the peaks in an NMR spectrum

Further Reading

Henzler-Wildman K, Kern D (2007) Dynamic personalities of proteins. Nature 450:964–972.

Levitt M (2001) Spin dynamics: basics of nuclear magnetic resonance, Chaps. 15 and 16. John Wiley & Sons, Inc., Chichester.

Mittermaier A, KaY LE (2006) New tools provide new insights in NMR studies of protein dynamics. Science 312:224–228.

Chapter 6
Protein Dynamics

Sometimes ignorance is not really such a bad thing.

Attila Szabo

Three final exams, two term papers, and an oral presentation—plus, you still need to clean your room and pack for a summer trip. The end of the semester is a blizzard of activity, pulling you in a hundred directions at once. It can be so overwhelming that you sometimes don't know where to begin. One of the best strategies for coping with the chaos is to make a "to-do list," start from the top of the list, and then attack each item one by one. Breaking down a jam-packed schedule like this creates order and organization, making the entropy more manageable and the tasks less formidable.

Protein motion and dynamics is a busy schedule on steroids—you've got molecules tumbling around like pasta in boiling water, loops swinging like vines on a windy lake, atoms vibrating around bonds like shocks in a car, and aromatic rings flipping around like pancakes on a griddle. Where in the heck do we begin to characterize this mess?

That's where NMR relaxation analysis comes to our rescue. Similar to a well-organized "to-do" list, NMR relaxation studies help us break the seemingly chaotic dynamics of proteins into a simple list of possible motional time scales. Then we simply go down the list, determine

M. Doucleff et al., *Pocket Guide to Biomolecular NMR*, 131
DOI 10.1007/978-3-642-16251-0_6, © Springer-Verlag Berlin Heidelberg 2011

the frequency of this motion, and see how much this component contributes to the nucleus's total relaxation (its T_1 and T_2 values). We keep adding more motion until all the relaxation is all accounted for. Let's see how it works!

6.1 Dynamics Analysis by NMR: Multi-channel Metronomes, Not a GPS

NMR is one of the few experimental techniques that provides details for the motion of individual atoms in large biomolecular molecules. Given the link between protein motion and protein function, it's no wonder that dynamic studies by NMR are an extremely hot topic in biochemistry. But when a structural biologist says, "I study protein dynamics by NMR," what exactly does he or she mean? What exactly is the nature of the information obtained?

A good analogy is a tracking device system, like a GPS. For example, if your parents want to monitor your "dynamics" at college, they could attach a GPS to your car (or even your favorite belt) and record everywhere you go. Eventually they could create a map of your daily motions, plotting where you eat, where you study, and where you go to relax.

It would be fantastic if we could do the same for a protein—attach a sub-microscopic GPS unit to a few atoms on a protein and track its movements as it progresses through its duties of catalyzing reactions and binding targets. Although this is a lofty goal of many NMR studies, this "geographical" information is not available from standard NMR dynamics experiments. Instead, NMR relaxation experiments give us two major types of information about the motion and dynamics of biomolecules:

(1) The *frequency of the motion*, or how rapidly the motion occurs, which is usually given as a *correlation time* or *constant*, τ. The correlation constant tells us how long it takes the motion to start

repeating itself, just like we learned about in Chap. 5, Sect. 5.6. Very rapid fluctuations (Fig. 5.12a) have small correlation times, and slow motions (Fig. 5.12b) have large correlation times (see the Mathematical sidebar below for more information on correlation times).

(2) The *amplitude of the motion*, or how far the atoms move from an average position during their dynamics. This is usually given in terms of an *order parameter*, S^2.

The coolest part about NMR relaxation studies is that even when an atom undergoes multiple types of motion—say its vibrating about a bond, moving within a flexible loop, and tumbling in solution with the rest of the protein—NMR experiments can tease out all these motions, determine how often each occurs (the correlation time) and, calculate how "big" the motion is (the order parameter). Let's begin with the fast motions and work our way up to the slower ones.

6.2 Elegant Simplicity: Lipari and Szabo Throw Out the Models

Attila Szabo, a senior research scientist at the National Institutes of Health, is as close to a "rock star" as you'll find in physical chemistry. Young scientists flock around Dr. Szabo at international conferences like biophysics paparazzi, requesting autographs of his quantum mechanics book and a 1982 paper in the *Journal of the American Chemical Society* (see References and Further Reading). Back home at NIH, Dr. Szabo is never shy to embrace the rebellious rock-n-roll lifestyle. Sporting the classic handlebar mustache, Dr. Szabo is infamous for sneaking in a smoke or two in his office, despite NIH's strict tobacco-free policy.

Why is this devilish theoretician the Keith Richards of biophysics? Over 25 years ago, Dr. Szabo and his graduate student, Giovanni Lipari, did the impossible—they made complicated science *simple*.

Lipari and Szabo realized that NMR spectroscopists were making protein dynamics more complicated than was required. At the time, spectroscopists were applying complex models of atomic motion to explain the decay of the NMR signal, the T_1 and T_2 curves. Lipari and Szabo realized that all these convoluted equations and simulations were completely unnecessary (and not always useful). In fact, models aren't needed at all! Instead, Lipari and Szabo showed that the fast motion of atoms (those on the picosecond to nanosecond time scale) is easily described by only three parameters:

(1) A *global rotational correlation time* (τ_c) that describes the *global* tumbling of the molecule in solution (Fig. 6.1a)—in other words, how long does it take for the protein to rotate around in solution like the pasta shell in boiling water, just as we learned in Chap. 5, Sect. 5.6. Lipari and Szabo originally called the global correlation time "τ_m," where the "m" stands for "molecular" because the τ_m is the same for the entire *molecule*. However, nowadays NMR spectroscopists also frequently call this term τ_c.
(2) A *local correlation time* (τ_e) that describes any fast motion (picosecond to nanosecond time scale) present at a specific location in the protein (Fig. 6.1b). For example, how quickly does the atom vibrate or move around *locally*. The "e" stands for "extra" because it represents motion *in addition* to the overall tumbling of the molecule. Unlike τ_c, this value can vary between different parts of the protein.
(3) An *order parameter* (S^2) that tells us the percentage of motion coming from the global tumbling compared to the local, "extra" motion. 1-S^2 gives us the percentage derived from the local fluctuations.

Because the values of these parameters don't depend on a model for the motion, Lipari and Szabo named the approach the "Model Free" analysis. Twenty-five years later, this simple method is employed in almost all NMR dynamics studies, and it is definitely the place to begin when analyzing molecular motion.

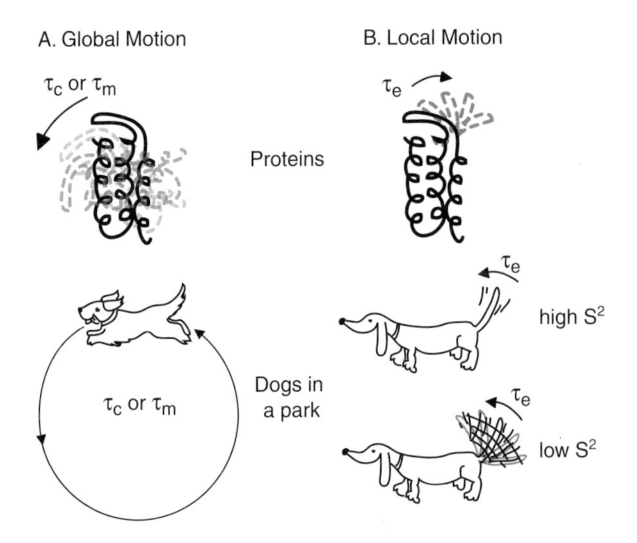

Fig. 6.1 (**a**) Proteins in solution and dogs in a park both have global motion with rotational correlation times of τ_c (or τ_m) that is the same for all parts of the molecule or the canine; (**b**) they also both have regions of local motion with correlation times of τ_e that are specific to only particular regions of the molecule or dog. This local motion can have a small amplitude (high S^2) or a large amplitude (low S^2)

6.3 Wagging Tails and Wiggling Bottoms: Local Versus Global Motion

To get more acquainted with the terminology used in Model Free analysis, let's look at a funny analogy. Have you ever watched a dog run around a park? Think of an especially wiggly one, like a portly Labrador retriever or extra-long German Shepherd. Similar to proteins in solution, a dog has multiple modes of motion, each with its amplitude and frequency. First, of course, is the dog's forward motion through space as it circles around the field (Fig. 6.1a, bottom). Just as the protein tumbles around in solution at a specific rate (Fig. 6.1a, top), the dog usually loops around the park at some characteristic "rotational correlation time." The quicker it runs, the faster its path in the field starts to repeat itself, and the smaller its correlation time, τ_c. In contrast, slower dogs take longer to loop around the park, and thus their rotational correlation times will be larger. This rotational correlation time is the same for every part of the doggy—his head and front legs circle the field just as fast as the tail and furry bottom. Thus, NMR spectroscopists call this rotational correlation time, the "overall or global correlation time."

Take a closer look at this frolicking canine, and you'll find many other types of motion present in only particular parts of the dog and independent of his global movement. For example, look at the rear section of the dog: many dogs wiggle their back-end and wag their tails as they trot along (Fig. 6.1b, bottom). This sash-saying has a characteristic correlation time that is completely independent of the overall correlation time around the field. Plus, this motion applies to only a small part of the dog—its back-end! For proteins, NMR spectroscopists call this "local" or "internal" motion because the movement is restricted to only certain regions or residues of the protein (Fig. 6.1b, top).

Smaller dogs usually jiggle their back-end much faster than larger dogs. Thus, the local correlation, τ_e, will be *smaller* than that for larger dogs because it takes less time for the faster wiggling to start to repeat itself than the slower wagging of heavier dogs. In contrast, the larger dogs tend to have *bigger* wiggles (Fig. 6.1b)—they usually move their tails and bottoms much farther from the center of their back than smaller dogs. In other words, the tail and back-end motion of bigger dogs has a larger *amplitude*, which contributes more to the total motion of the dog's rear-end than the tail motion of smaller dogs.

NMR spectroscopists designate the amplitude of this local wiggling by what's called the "order parameter," or S^2. Specifically, the order parameter tells us the probability of finding the dog's tail in the center of its body. If the tail doesn't have any local motion at all, then the order parameter is 1.0 because there is a 100% chance that the tail will be sticking straight out. Now if the tail is whipping violently from one side to the other, the order parameter will be much lower, like 0.5, because the chance of finding the tail in the center is significantly lower, say 50%. Thus, the order parameter gives us an estimate of the rigidity of the dog's back-end. *The smaller the* S^2*, the more violent the wiggles; the higher the* S^2*, the more demure the motion* (Fig. 6.1b, bottom).

Be sure to remember that the order parameter is a probability so it is always between 0 and 1.0, or 0 and 100%, with 1.0 describing a completely stiff region and 0 being super wiggly in all directions (i.e., the wiggles are *isotropic* or the same in all directions).

Proteins are just like that energetic dog at the park: they have global motion that is the same for all parts of the protein (Fig. 6.1a, top) and local motions that are specific to certain regions of the protein (Fig. 6.1b, top). Each one of these motions creates fluctuating magnetic fields and therefore, causes the NMR signal to decrease. Our job is to figure out the speed of each motion and to determine how much each type of motion contributes to the overall relaxation of an atom's NMR signal. That's where the Model Free analysis comes into play.

6.4 Measuring Fast Motion: *Model Free* Analysis

The Model Free analysis formulated by Lipari and Szabo is incredibly easy. It provides simple equations that relate the T_1 and T_2 relaxation times to the three parameters described above, τ_c, S^2, and τ_e, which you can think of as a dynamics "to-do" list (Fig. 6.2). First we measure the T_1 and T_2 values to determine the total amount of relaxation present at the atom.[1] Then we simply go down the list of parameters, calculate its value, and add that type of fluctuation to the total motion creating the relaxation. If this motion isn't sufficient to account for the all the relaxation observed in the NMR experiments (in other words, there is still some "relaxation" leftover), then we go to the next item on the list

[1] Measuring T_1 and T_2 is not too complicated. You essentially record multiple ^1H-^{15}N HSQC spectra with an increasing delay inserted in the sequence to allow either transverse or longitudinal relaxation. However, for the sake of clarity, simplicity, and length in this book, we direct the reader to pp. 315–325 in *Spin Dynamics* by Malcolm H. Levitt (Wiley, 2002).

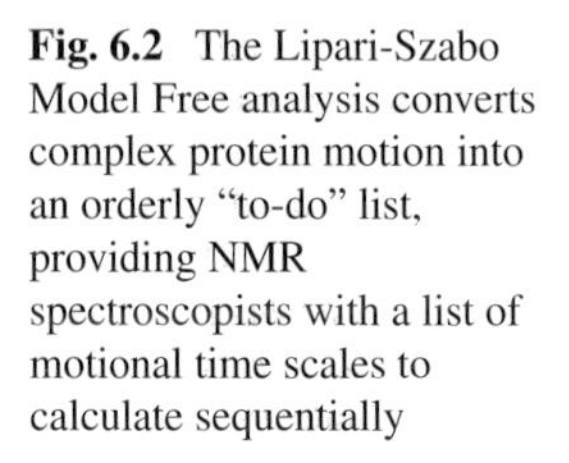

Fig. 6.2 The Lipari-Szabo Model Free analysis converts complex protein motion into an orderly "to-do" list, providing NMR spectroscopists with a list of motional time scales to calculate sequentially

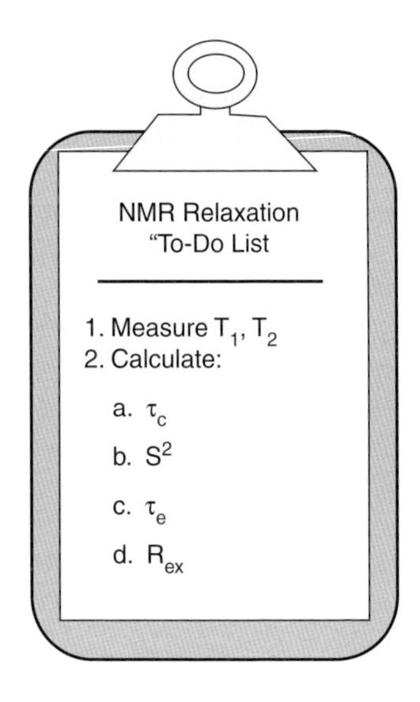

and repeat the process. That doesn't sound too bad, does it? So let's start at the top of the list with the global tumbling or τ_c (Fig. 6.2):

No matter the shape, size, or function, all proteins tumble around in solution. Therefore, all proteins have an overall rotational correlation time (just like all doggies have a correlation time for running around the park). In fact, *the global tumbling is the major contributor to both the* T_1 *and the* T_2 relaxation of the amide hydrogens. So we start there: Determine τ_c for each amide hydrogen.

It turns out that calculating τ_c is quite easy. When the overall tumbling of the protein is the only significant motion present at an atom (in other words, the local motion doesn't contribute significantly to the relaxation of the hydrogen), the global rotational correlation time is approximately proportional to the ratio of T_2/T_1—amazingly, simple!

So we first estimate as $\tau_c \alpha T_2/T_1$, and then check to see if this type of motion is sufficient to account for our T_1 and T_2 values:

(a) If it is, then we are finished for that atom! We know that the only significant motion present at that atom is your run-of-the-mill tumbling in solution and not much else.
(b) If the global tumbling isn't enough to account for our T_1 and T_2 values for this atom, then we need to add more motion to the system.

Before we do that, let's talk a bit more about how we know if this motion "accounts" entirely for the relaxation plot. We use two methods:

1. We use the equations provided by Lipari and Szabo to estimate the order parameter S^2 (see Mathematical Sidebar 6.1 for a glimpse of the math behind Model Free analysis). Remember that the order parameter tells us the percentage of motion coming from the overall tumbling of the molecule. So if S^2 is 1.0 or near 1.0, then we know that majority of the atom's motion is due solely to the global tumbling in solution. For example, check out the relaxation data in Fig. 6.3; residues 83–89 (highlighted in gray) have an order parameter of ~0.7 or less, and thus this region or the protein must have some local motion not present in the other residues (besides the N-terminal and C-terminal ends). In contrast, residues 43–55 have order parameters from 0.8 to 1.0, so we know that the majority of the motion present at these residues comes solely from the overall tumbling of the molecule.
2. We check the τ_c (α T_2/T_1) values of all the other residues. Remember that the overall rotational correlation should be the same for all residues in the protein (Fig. 6.1a). If the τ_c value for a residue doesn't match the others, then additional motion must be present at this residue. For these residues, we need to add local motion. Look again at the data in Fig. 6.3; across the entire protein, the τ_c value is scattered about ~7.8 ns (gray line), except for the region highlighted in gray. Residues in this region must have *local* motions that skew the calculated τ_c value.

According to the Model Free analysis (Fig. 6.2), we next add to the system a very fast motional component—dynamics that occur on the

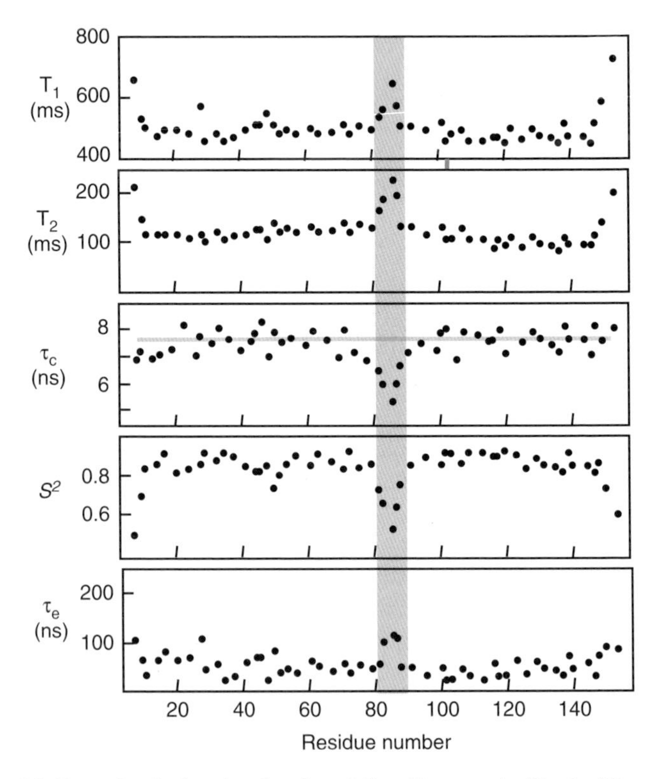

Fig. 6.3 Example of relaxation data for a 148-residue protein. The flexible region in the center of the protein is shaded in gray, and the τ_c, approximated for the whole protein, is designated by the gray line in the middle panel

picosecond to nanosecond time. Again, we use the equations provided Lipari and Szabo to estimate the correlation time, τ_e, of this fast, localized motion and then recalculate the order parameter S^2. Remember that S^2-1 tells us how much of the total motion comes from these regional fluctuations, but it also gives us an idea of how far the atom vibrates away from its average position in the protein (see the Mathematical Sidebar in this chapter for more information about Lipari–Szabo's Model Free equations).

In general, if the S^2 *is smaller than 0.8 than the residue contains significant amount of local motion. In contrast, an* S^2 *value larger than 0.8 indicates very little fast motion beyond the overall tumbling of the molecule.*

Okay, now is where the Model Free analysis gets a bit tricky. What if the T_1 and T_2 values are still not satisfied by τ_c, τ_e, and S^2? Well, we have a few options. First, we can try to add another fast motion component with its own correlation constant (τ_{e2}), telling us the time scale of the motion, and its own order parameter (${S_2}^2$), supplying the percentage that this motion contributes to the total motion. This approach is called the *extended model free*, and it works well when we have two unique types of motion on the picosecond to nanosecond time scale.

But what if we also have slower motion? The Model Free analysis does a great job of finding the frequencies and amplitudes for motions that are fast (compared to the chemical shift time scale), but it doesn't give us much information for the dynamics at the intermediate time scales, like the microsecond to millisecond time scale—the motions that broaden and distort our NMR peaks (Fig. 5.9b). For these slower motions, the Model Free analysis simply lumps them all together into one rate called the "exchange factor" or R_{ex}. The R_{ex} is essentially a "fudge factor" saying, "Ok, there is definitely some slower motion present at this atom, and it occurs at approximately the rate R_{ex}." Unfortunately, though, this value is not very accurate or informative. Why?

Remember from Chap. 5, when motion occurs in the intermediate time regime (microsecond to millisecond) the T_1 and T_2 values depend not only the rate of the motion but also on the difference(s) in the ringing frequency of the atom at the locations visited during the motion. So now we have another "chicken-and-the-egg" problem—we need the ringing frequencies to get the rate of the slow motion, but we need the rate to get the ringing frequencies. How in the world do we attack this problem?

Fortunately, there is a cool trick that can tease out both the ringing frequencies (or chemical shifts) and the exchange rates at the intermediate time regime. It is called *CPMG relaxation dispersion analysis*, and although it has a rather complicated name, it is not difficult to understand, especially once you master the concept of a *refocusing pulse*. So let's do that first, and then we'll dive into CPMG relaxation dispersion.

Mathematical Sidebar 6.1: Correlation Functions and Model Free

Not feeling so satisfied by that highly descriptive explanation of the Model Free? Craving a few Greek symbols and operators for correlation times? No problem. Here's a double-hitter mathematical sidebar for both correlation functions and the Model Free approach. Actually, correlation constants are so important in protein dynamics and NMR spectroscopy that they deserve a more detailed description, even a few equations. Don't be afraid, it isn't very complicated.

In the previous chapters (Chaps. 4 and 5), we learned that all molecular motion creates fluctuating magnetic fields around a nucleus. These oscillating fields cause an atom's ringing frequency (and chemical shift) to change, leading to dephasing and a loss in the total NMR signal. If we graph the strength of the total magnetic field felt by an atom over time, it would look like a Mexican jumping bean, going up and down randomly around the average value, which is zero in the example shown in Fig. M6.1a.

Unlike our beautiful NMR signal, which consists of only sine curves with well-defined frequencies, these fluctuations are quite complicated and random. Indeed, we can't define specific frequencies for these oscillations, like we do for the ringing nuclei. How on Earth do we measure or quantify the speed of these fluctuations?

Fig. M6.1 Plots of (**a**) the strength of the magnetic field at an atom in the protein over time ($B(t)$); (**b**) a plot of $B(t = 0) \times B(\tau)$ versus the time interval τ. (**c**) $B(t_A) \times B(t_A + \tau)$ versus the time interval τ; the point t_A is shown in panel A. (**d**) $B(t_B) \times B(t_B + \tau)$ versus the time interval τ, the point t_B is shown in panel A. (**e**) the average of all the $B(t) \times B(t + \tau)$ curves, $\langle B(t) \times B(t + \tau)\rangle$, versus the time interval τ

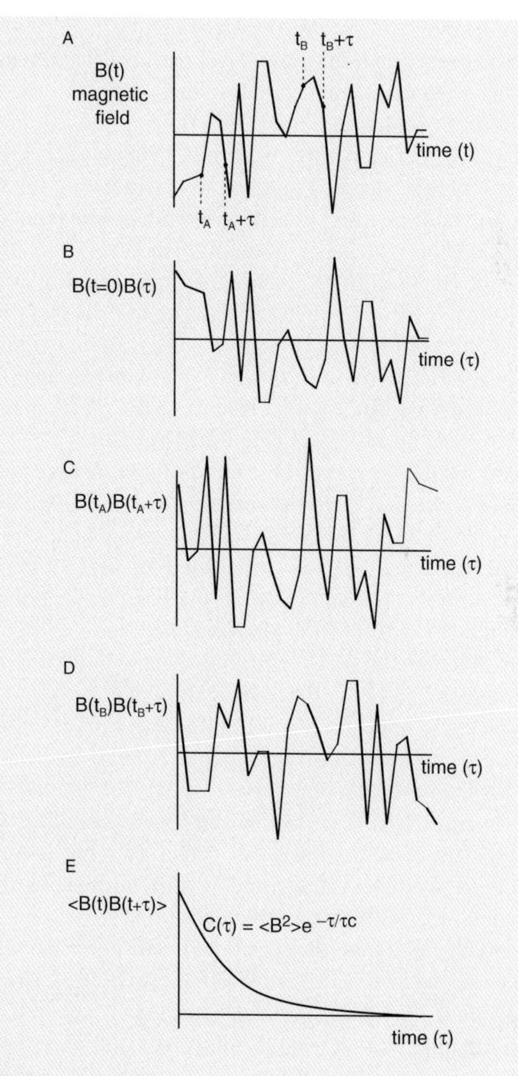

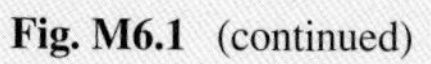

Fig. M6.1 (continued)

Again a great physical chemist comes to our rescue and provides us with an elegant way to define precisely the speed of a randomly oscillating curve. This time it is Lars Onsager, a Norwegian-American who won the 1968 Nobel Prize in Chemistry. Onsager decided to take the first point on the curve, $B(t = 0)$, and multiply this value by every other time point on the curve, say at time τ. When τ is small, then $B(t = 0) \times B(\tau)$ is large because $B(\tau)$ and $B(t = 0)$ have the same sign (Fig. M6.1b). As τ gets bigger, $B(t = 0) \times B(\tau)$ decreases and crosses through zero when $B(t = 0)$ and $B(\tau)$ begin having different signs. But eventually the $B(t = 0) \times B(\tau)$ curve starts to rise again, and, indeed, a graph of $B(t = 0)$ x $B(\tau)\tau$ looks quite similar to $B(t)$—fluctuating up and down (Fig. M6.1b).

OK, now what if we create these curves for *more* time points t? In other words, we pick anywhere on the $B(t)$ curve (Fig. M6.1a), say $B(t_A)$, and multiply this value by $B(t_A + \tau)$, where τ is some interval of time *after* t_A. Now when we plot $B(t_A) \times B(t_A + \tau)$ (Fig. M6.1c), we get a curve that, at the beginning, looks very similar to the $B(t = 0) \times B(\tau)$ plot (they both have a large positive value), but then the $B(t_A) \times B(t_A + \tau)$ curve quickly diverges from the first plot from the $B(t = 0) \times B(\tau)$ plot.

If we pick any other time point on the $B(t)$ curve, say $B(t_B)$, we see a very similar result (Fig. M6.1d): A large positive value initially but then a fluctuating curve distinct from the other two graphs as τ gets larger (Fig. M6.1b and M6.1c). In other words, these curves are *correlated* at small values of τ but loose their correlation as τ grows. How quickly this correlation decays depends on how quickly $B(t)$ oscillates!

To quantify this loss of correlation, Onsager had a brilliant idea: he decided to *average* all these $B(t) \times B(t + \tau)$ curves together. The result is rather amazing: It converts these random oscillating curves into a silky, smooth exponentially decaying curve that is quite similar to the exponentially decay we see in the NMR signal (Fig. M6.1e). This exponential curve is called the *autocorrelation function* or $C(\tau)$.

The nice part of the autocorrelation function is that we can fit it to a simple exponential function and determine the decay rate constant (R). The faster the fluctuations in the original $B(t)$ curve, the faster the average of $B(t)B(t+\tau)$ goes to zero, and the larger the decay rate constant. Thus, the decay rate of the correlation function gives us a precise value for how quickly the magnetic field fluctuates. Very nice!

$$C(\tau) = \text{average}(B(t) \times B(t+\tau)) \text{ proportional to } e^{-Rt} = e^{-t/t_C}$$

Unfortunately, though, NMR spectroscopists make it slightly more complicated than this. For historical reasons, they don't ever talk about the correlation decay *rate*, R. Instead, they always use $1/R$, called the correlation *constant* or τ_C. So the correlation constant, τ_C, tells us approximately how long it takes for the fluctuating magnetic field to pass through the average value.

Lipari and Szabo were the first to recognize that the T_1 and T_2 values measured in NMR experiments can easily be defined in terms of simple correlation functions, $C(\tau)$s, for the fluctuating magnetic fields created by molecular motion. Since we don't need to know anything about the nature of the atomic motion to estimate the speed of the fluctuating field, they called it the "model-free" approach.

But Lipari and Szabo recognized that using correlation functions to interpret NMR data is cooler than this—they realized that this method easily allowed them to dissect out different motional time scales from the fluctuating fields. When a curve contains multiple speeds of oscillations (that are independent of each other), say fluctuations due to local bond vibrations and global molecular tumbling, then the total correlation function is simply the product of the individual correlation functions, each with their own correlation constant, τ_c and τ_e:

$$C_{Total}(\tau) = C_{Global}(t)C_{Local}(t) = S^2\, e^{-t/t_C} + (1 - S^2)\, e^{-t/t_e}$$

And, then the order parameter, S^2, simply tells us how much the global correlation contributes to the total correlation, $C(\tau)$.

6.5 Changing Directions on the Track: Refocusing Pulses

To understand a refocusing pulse, you need to remember two major concepts about NMR:

1. When an atom rings, its small magnetic needle in the nucleus rotates around the *z*-axis, like it is sitting on an old-school record turntable (Chap. 4, Sect. 4.3)
2. We don't record the ringing of one atom in the sample, but rather one specific *type* of atom with millions-upon-millions of copies all ringing together in unison, or "in phase." If their ringing becomes out-of-sync (i.e., all the little magnetic needles start to spread out in the *xy*-plane; see Chap. 4, Sect. 4.5), the NMR signal decreases and the peak in the NMR spectrum broadens (Fig. 4.18).

With these ideas in mind, let's return to "NMR track-and-field" to learn about refocusing pulses. Imagine you have two runners standing at the starting line on a track (Fig. 6.4a). The gun goes off, and they both start running counterclockwise. But the race isn't evenly matched. One runner is an elite short-distance track-and-field star that easily maintains a 5 minutes-per-mile pace. In contrast, the other runner is an average high

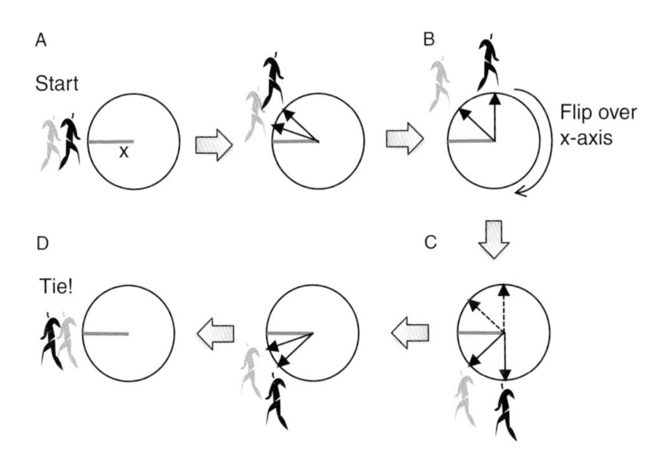

Fig. 6.4 (**a**) Two runners with different paces start a race at the *x*-axis. (**b**) Twenty seconds into the race, (**c**) the runners are magically "flipped" to the opposite side of the track. (**d**) If the runners maintain their exact paces, they reach the finishing line (the *x*-axis) at exactly the same time

school athlete that immediately settles into a strict 7 minute-per-mile pace.

So what does the race look like after 20 seconds into the race? That's easy, right? The faster runner is ahead (Fig. 6.4b). Okay, but now what happens if the runners are magically flipped to the opposite side of the track; instead of being near the starting line, they are suddenly near the finish line (Fig. 6.4c). Assuming each runner maintain his pace, who wins the race?

Surprisingly, it's a perfect tie (Fig. 6.4d)! Flipping the runners shifts their order—the slower runner is now closer to the finish line than the faster one. Because the runners maintain their individual paces, the faster one catches up with the slower one exactly at the finish line, creating a flawless photo-finish (Fig. 6.4d)!

The little magnets in nuclei act the same way. When we ring the nuclei, they all start at the same place in the *xy*-plane, producing a nice strong NMR signal (Fig. 6.5a). However, as they "run" around the

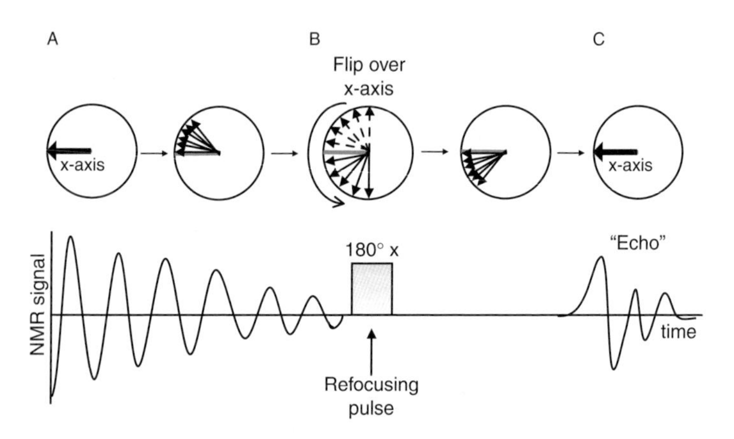

Fig. 6.5 (**a**) As the nuclei rotate around the z-axis, they fall "out-of-synchrony," causing the NMR signal to decrease (bottom). (**b**) A refocusing (or 180°-π) pulse flips the nuclei over the x-axis (**c**) If the nuclei don't change their ringing frequencies, then they reach the x-axis at exactly the same time, recreating the NMR signal and producing an NMR "echo."

z-axis, atoms with faster ringing frequencies will quickly get ahead, and the slower ones will fall behind, making our NMR signal fade away. But if we flip the nuclei across the x-axis (Fig. 6.5b), then viola! They all come back together and recreate the original NMR signal at the x-axis (Fig. 6.5c). This "flipping" around is called a *refocusing pulse, a π-pulse* or a *spin-echo* because the signal vanishes but then reappears after a short delay. When you see any of these names, just remember that this is a trick for bringing the nuclei back into synchrony and for "correcting" any dephasing due to differences in ringing frequency.

6.6 Measuring Intermediate Motion: CPMG Relaxation Dispersion Analysis

The key point about refocusing pulses is that they work *only if the nuclei maintain their exact ringing frequency*. Think back to the runners in the previous section (Fig. 6.4). For the faster runner to catch the slower one at the finish line (Fig. 6.4d), she has to maintain exactly the same pace as she did at the beginning of the race—if she speeds up, then she'll pass the slower runner before the finish line; if she decreases her speed, then she won't have enough time to catch the runner.

The same goes for the atoms—if the atom's ringing frequency changes before or after the refocusing pulse is applied, their spinning will still diverge, and we won't see the "echo" that appears after the 180° refocusing pulse (Fig. 6.5c). Although this seems like a severe limitation of refocusing pulses, in fact, it is the key attribute that makes them useful for analyzing dynamics on the intermediate time scale (microsecond to millisecond time scale).

The idea is rather simple: instead of using only one refocusing pulse, we apply many pulses one after another. If the time between the pulses is small enough to flip the nuclei before they have a chance to switch locations and ringing frequency, then the refocusing pulses keep the nuclei in synchrony, and we see a nice strong, narrow peak in the NMR spectrum (Fig. 6.6c). However, if the time between the refocusing pulses is too long, and the nuclei have the opportunity to switch locations between pulses, then the refocusing pulses don't work, and we see the same broad, smeared-out peak in NMR spectrum that we obtain without the refocusing pulses (Fig. 6.6a). Therefore, *we can figure out how quickly the atoms exchange between two locations by determining the pulsing frequency required to effectively refocus the nuclei and sharpen the NMR peak.*

Figure 6.6 nicely illustrates the details of the CPMG relaxation dispersion analysis. In panels a–c, each atom is represented by a line, and its ringing frequency is given by the slope of the line. Each refocusing pulse (black rectangles) switches the direction of the slope. The lines diverge as the atoms switch their ringing frequencies, leading to a broad peak in the NMR spectrum and a short T_2 value (panel d).

In CPMG analysis, we begin by applying only a few refocusing pulses that are spaced rather far apart (Fig. 6.6a). In this case, the refocusing pulses can't "keep up" with the changing frequencies, so the

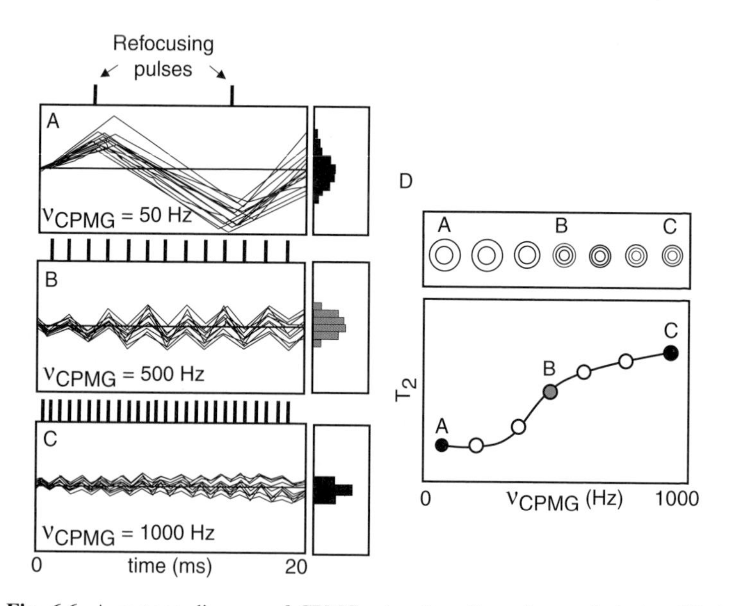

Fig. 6.6 A cartoon diagram of CPMG relaxation dispersion analysis (modified from Mittermaier and Kay (2009) Trends Biochem Sci 34:601–611). (**a–c**) Each atom is represented by a line, and its ringing frequency is given by the slope of the line; refocusing pulses are shown as black rectangles, and histograms of the ringing frequencies after the CPMG period are shown at the right. (**d**) Cartoon representation of the peak in the NMR spectrum as the frequency of refocusing pulses (ν_{CPMG}) increases (top); plot of T_2 versus ν_{CPMG} (bottom). The letters in panel D refer to the experiments in panels **a–c**

lines continue to diverge over time, and the peak in the NMR spectrum is broad (peak a in Fig. 6.6d).

When the refocusing pulses are applied at a rate that is similar to the exchange rate of the atoms (Fig. 6.6b), the atoms don't have a chance to change their location (or ringing frequency) between pulses, so the NMR signal stays strong over time, and the peak begins to narrow (peak b in Fig. 6.6d).

With even more refocusing pulses, spaced even closer together in time (Fig. 6.6c), the nuclei ring in almost perfect synchrony, resulting in an even stronger NMR signal and skinnier peak (peak c in Fig. 6.6d). Now we've almost completely "corrected" for the protein motion occurring in the intermediate time regime—it's like we've "thrown a wrench" into the protein's dynamics on the millisecond to second time scale!

If we plot the T_2 value versus the frequency of the refocusing pulses (ν_{CPMG}), we get a curve like the one shown in Fig. 6.6d (bottom). The shape of this curve depends directly on the exchange rate of the atoms (how quickly they switch between the two locations), the difference in ringing frequency between the two locations, and the percentage of atoms in the two locations. By fitting this curve to a set of sophisticated equations, we can determine the values all of these parameter. Take that "chicken-or-the-egg" problem!

One last note about CPMG relaxation dispersion analysis: this technique assumes that the atom is switching between only two different locations. If the motion is more complicated, then CPMG relaxation dispersion analysis won't be as helpful, and you'll have to wait a few years for a new NMR relaxation method to tackle these more complex situations.

6.7 Measuring Slow Motion: Z-Exchange Spectroscopy

Okay, so we characterize fast motion with Model Free analysis, and intermediate motion with CPMG relaxation dispersion analysis (Fig. 6.6), but what about slow motion?

Well, if the atoms are moving *too slowly*, like only a few times per minute, then NMR spectroscopy isn't helpful because our NMR signal rarely lasts longer than a second (Fig. 5.7). However, if the motion's rate sits right in the a sweet spot between 1 and 10 times per second, then we can usually use a technique called *Z-exchange spectroscopy* to nail down the rate of motion.

To see how Z-exchange spectroscopy works, let's go back to the situation where an amide hydrogen atom is exchanging between two locations, A and B, and its chemical shift is also switching between 7.0 and 8.0 ppm, respectively (Fig. 6.7a). Now let's say that the nitrogen covalently bonded to the hydrogen also changes its chemical shift in the two locations: in location A it rings at 120 ppm, but in location B it rings at 110 ppm. If we run a regular ^{1}H-^{15}N 2D-HSQC experiment on this system, what does the spectrum look like? Well, if the exchange between the two locations is slow enough, which is true in this case ($R = 2\ s^{-1}$), then we will see two peaks in the spectrum: one for the hydrogen in location A and one for the two atoms in location B (Fig. 6.7b).

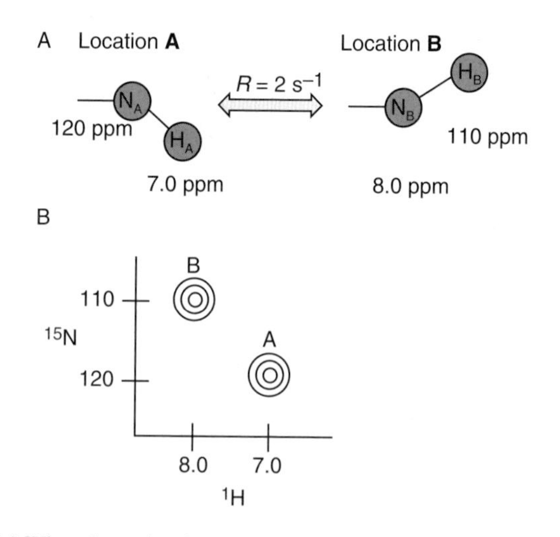

Fig. 6.7 (**a**) When the molecule moves from location a to location b, the chemical shift for the nitrogen atom switches from 120 to 110 ppm, and the chemical shift for the hydrogen atom switches from 7.0 to 8.0 ppm. (**b**) When this conformational change occurs infrequently compared to the chemical shift time scale, we see two peaks in the ^{1}H-^{15}N 2D-HSQC spectrum

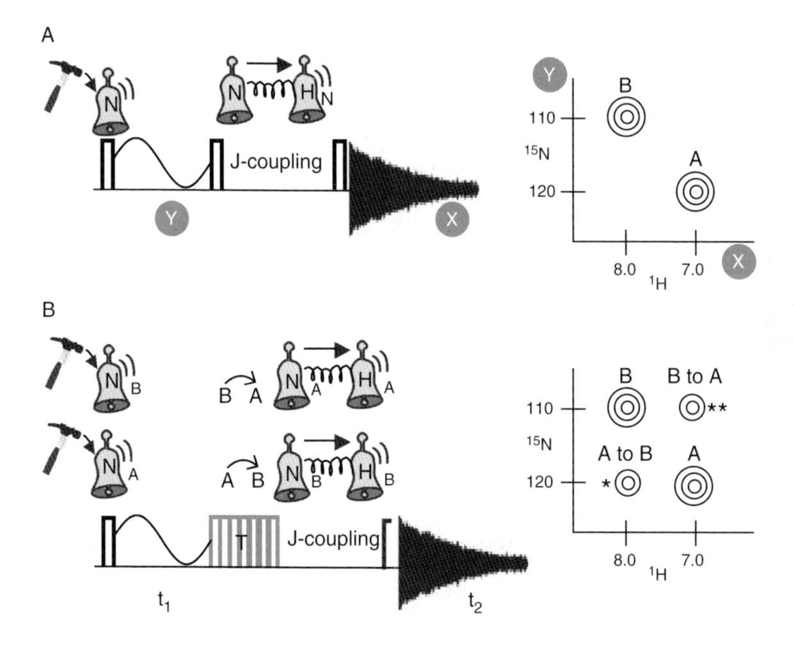

Fig. 6.8 (**a**) In the standard ^{1}H-^{15}N 2D-HSQC experiment, we ring the nitrogen atom and record its ringing frequency; let it ring the attached hydrogen atom; and then record the hydrogen's ringing frequency (right). For the system shown in Fig. 6.7, this produces two peaks in the spectrum (right). (**b**) In Z-exchange spectroscopy, we insert a short delay in the experiment (the *T*1 period, shown as a striped rectangle). If a significant portion of the molecules switches configuration during this period, then we see four peaks in the spectrum

Why is this true? Think back to how the HSQC experiment works (Chap. 2, Sect. 2.3 and Fig. 2.10) (Fig. 6.8a): We ring the nitrogen and record its signal; let the nitrogen ring the attached hydrogen through J-coupling; and then record the hydrogen's signal. As a result, the hydrogen atom broadcasts the ringing frequency of the nitrogen atom (Fig. 6.8a, left; Chap. 2, Sect. 2.5). If the system doesn't change locations during the experiment, then the hydrogen in location A (ringing at 7.0 ppm) broadcasts the signal for the nitrogen in location A (ringing at 120 ppm), and the hydrogen in location B (ringing at 8.0 ppm) broadcasts the signal for the nitrogen in location B (ringing at 110 ppm). The result is a spectrum with two distinct peaks (Fig. 6.8a, right): one at $x = 7.0$ ppm, $y = 120$ ppm; and one at $x = 8.0$ ppm, $y = 110$ ppm.

Now, let's throw a twist into the HSQC experiment—let's add a small delay right before the nitrogen rings the bonded hydrogen via J-coupling (Fig. 6.8b, striped rectangle). If a significant percentage of the hydrogen and nitrogen atoms move from location A to B and vice versa during this period, then guess what? We get four peaks in the HSQC spectrum (Fig. 6.8b, right)! We see the same ones as before, but we also find two more peaks at $x = 7.0$ ppm, $y = 110$ ppm and $x=$ 8.0 ppm, $y = 120$ ppm.

The first extra cross-peak at 8.0 and 120 ppm (* in Fig. 6.8b, right) comes from molecules that are in location A while the nitrogen rings but in location B when the hydrogen rings. The second extra peak at 7.0 and 110 ppm (** in Fig. 6.8b, right) represents molecules in location B while the nitrogen is ringing but in location A while the hydrogen rings.

In practice, spectroscopists usually run the Z-exchange experiment multiple times, each time varying the length of the delay (called the *T_1 period*). As the delay gets longer, the intensities of the extra cross-peaks increase while the intensities of the original peaks decrease. The exchange rate between locations A and B (Fig. 6.7a) is easily calculated from the rate at which the two new cross-peaks grow and the two original peaks shrink. This is how NMR spectroscopists measure the exchange rate via Z-exchange spectroscopy—pretty cool!

6.8 Measuring Motion Summary

In this chapter, we've learned about three NMR experiments for characterizing the molecular motion on three different time scales: (1) Fast motion on the picosecond to nanosecond time scale; (2) Intermediate

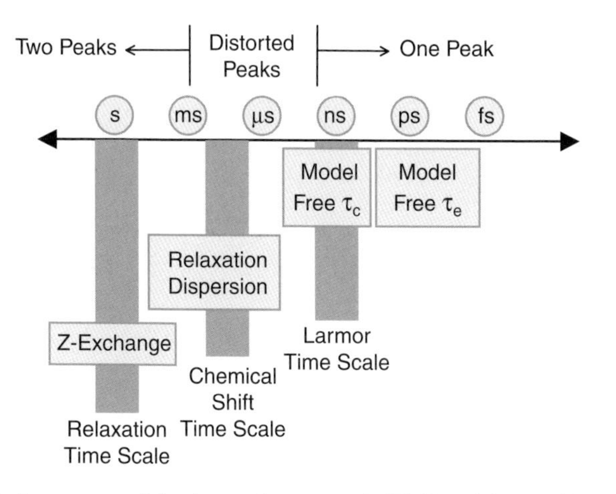

Fig. 6.9 A summary of the time scales present in NMR and the specific methods to characterize each one. The top of the diagram indicates how dynamics on these time scales affects the peaks in an NMR spectrum

motion on the microsecond to millisecond time scale; and (3) Slow motion on the microsecond to millisecond time scale. To help you remember it all, Fig. 6.9 summarizes which experiments go with each motional time scale.

References and Further Reading

Barbato G, Ikura M, Kay LE, Pastor RW, Bax A (1992) Backbone dynamics of calmodulin studied by ^{15}N relaxation using inverse detected two-dimensional NMR spectroscopy. *Biochemistry* 31:5269–5278.

Cavanagh J, Fairbrother WJ, Palmer AGIII, Rance M, Skeleton NJ (2007) Protein NMR spectroscopy: principles and practice, 2nd edn., Chap. 8. Academic Press, Amerstdam.

Clore GM, Szabo A, Bax A, Kay LE, Driscoll PC, Gronenborn AM (1990) Deviations from the simple two-parameter model-free approach to the interpretation of nitrogen-15 nuclear magnetic resonance of proteins. *J Am Chem Soc* 112:4989–4991.

Lipari G, Szabo A (1982) Model-free approach to the interpretation of nuclear magnetic resonance relaxation in macromolecules 1. Theory and range of validity. *J Am Chem Soc* 104:4546–4559.

Mittermaier A, Kay LE (2006) New tools provide new insights in NMR studies of protein dynamics. *Science* 312:224–228.

Mittermaier A, Kay LE (2009) Observing biological dynamics at atomic resolution using NMR. *Trends Biochem Sci* 34:601–611.

Neudecker P, Lundström P, Kay LE (2009) Relaxation dispersion NMR spectroscopy as a tool for detailed studies of protein folding. *Biophys J* 96:2045–2054.

Index

M. Doucleff et al., *Pocket Guide to Biomolecular NMR*,
DOI 10.1007/978-3-642-16251-0,

X

Z

Printing: Ten Brink, Meppel, The Netherlands
Binding: Stürtz, Würzburg, Germany